Theodore Gill

Arrangement of the Families of Mammals

Theodore Gill

Arrangement of the Families of Mammals

ISBN/EAN: 9783337864965

Printed in Europe, USA, Canada, Australia, Japan

Cover: Foto ©berggeist007 / pixelio.de

More available books at **www.hansebooks.com**

SMITHSONIAN MISCELLANEOUS COLLECTIONS.

—————— 230 ——————

ARRANGEMENT

OF THE

FAMILIES OF MAMMALS.

WITH ANALYTICAL TABLES.

PREPARED FOR THE SMITHSONIAN INSTITUTION.

BY

THEODORE GILL, M.D., Ph.D.

WASHINGTON:

PUBLISHED BY THE SMITHSONIAN INSTITUTION.

NOVEMBER, 1872.

ADVERTISEMENT.

The following list of families of Mammals, with analytical tables, has been prepared by Dr. Theodore Gill, at the request of the Smithsonian Institution, to serve as a basis for the arrangement of the collection of Mammals in the National Museum; and as frequent applications for such a list have been received by the Institution, it has been thought advisable to publish it for more extended use. In provisionally adopting this system for the purpose mentioned, the Institution, in accordance with its custom, disclaims all responsibility for any of the hypothetical views upon which it may be based.

JOSEPH HENRY,
Secretary, S. I.

Smithsonian Institution,
Washington, October, 1872.

(iii)

CONTENTS.

I. List of Families* (including references to synoptical tables) . . . 1–27

Sub-Class (Eutheria) Placentalia *s.* Monodelphia (1–121) . . 1, 45, 46
  Super-Order Educabilia (1–73) 1, 45, 46
    Order 1. Primates (1–8) 1, 47, 50
      Sub-Order Anthropoidea (1–5) . . . . 1, 50, bis
        "   Prosimiæ (6–8) 2, 50, 54
    Order 2. Feræ (9–27) 3, 47, 56
      Sub-Order Fissipedia (9–24) . . . . 3, 56, bis
        "   Pinnipedia (25–27) . . . 7, 56, 68
    Order 3. Ungulata (28–54) 8, 47, 70
      Sub-Order Artiodactyli (28–45) . . . 8, 70, 71
        "   Perissodactyli (46–54) . . . 11, 71, 84
    Order 4. Toxodontia (55–56) 13, 48, 89
    Order 5. Hyracoidea (57) 13, 48, 89
    Order 6. Proboscidea (58–59) 13, 48, 89
        Diverging (Educabilian) series.
    Order 7. Sirenia (60–63) 13, 48, 91
    Order 8. Cete (64–73) 14, 49, 92
      Sub-Order Zeuglodontia (64–65) . . . 14, 92, 93
        "   Denticete (66–71) 14, 92, 93
        "   Mysticete (72–73) 15, 93, 97
  Super-Order Ineducabilia (74–121) 16, 49
    Order 9. Chiroptera (74–82) 16, 49
      Sub-Order Animalivora (74–81) . . . 16
        "   . Frugivora (82) 18
    Order 10. Insectivora (83–92) 18, 49
      Sub-Order Dermoptera (83) 18
        "   Insectivora Vera (84–92) . . . 18
    Order 11. Glires (93–112) 20, 50
      Sub-Order Simplicidentati (93–110) . . 20
        "   Duplicidentati (111–112) . . 23
    Order 12. Bruta (113–121) 23, 50
      Sub-Order Vermilinguia (113) . . . 23
        Squamata (114) 23

* The numbers inclosed within parentheses refer to the families.

(v)

Sub-Order Fodientia (115) 23
" Tardigrada (116-117) 24
" Loricata (118-120) 24
Bruta incertæ sedis (121) . . . 25
Sub-Class (Eutheria) Didelphia (122-134) 25, 46
Order 13. Marsupialia (122-134) 25
Sub-Order Rhizophaga (122) 25
" Syndactyli (123-129) 26
" Dasyuromorphia (130-131) . . . 26
" Didelphimorphia (132) 26
Marsupialia incertæ sedis (133-134) . 27
Sub-Class (Prototheria) Ornithodelphia (135-136) . . . 27, 46
Order Monotremata (135-136) 27
Sub-Order Tachyglossa (135) 27
" Platypoda (136) 27

II. LIST OF AUTHORS REFERRED TO 31-41

III. SYNOPTICAL TABLES OF CHARACTER OF THE SUBDIVISIONS OF MAMMALS, WITH A
CATALOGUE OF THE GENERA 43-98

ARRANGEMENT

OF

FAMILIES AND SUB-FAMILIES

The present portion of the "Arrangement of the Families of Mammals" is issued in advance of the entire work. The completion is delayed on account of the inability of the author to consult certain works and examine the skeletons of several forms, but the remainder will be issued as soon as it can be prepared. Most of the pages now published have been stereotyped for more than a year, as will be seen from the dates at the bottom of the signatures.

SUB-ORDER ANTHROPOIDEA.

(*Bimana.*)

1. Hominidae   = Anthropini, Huxl., M. T. & G., 1864, i, 153.

(*Simiae.*)

(*Simiae catarrhinae.*)

2. Simiidae   = Anthropomorpha, Huxl., M. T. & G. 1864, i, 648.

vi

Sub-Order Fodientia (115) 23
    "   Tardigrada (116–117) 2ᴬ
    "   Loricata (118–120) 24
        Bruta incertæ sedis (121) . . . 25
Sub-Class (Eutheria) Didelphia (122–134) 25, 46
    Order 13. Marsupialia (122–134) 25
    Sub-Order Rhizophaga (122) 25
        "   Syndactyli (123–129) 26
        "   Dasyuromorphia (130–131) . . . 26
        "   Didelphimorphia (132) 26
        Marsupialia incertæ sedis (133–134) . 27

ARRANGEMENT

OF

FAMILIES AND SUB-FAMILIES OF MAMMALS.

[Adopted provisionally by the Smithsonian Institution.]

N. B.—The Fossil Families are indicated by Italics.

CLASS A.—MAMMALIA.

SUB-CLASS PLACENTALIA.

SUPER-ORDER EDUCABILIA.

(GYRENCEPHALA = MEGASTHENA + ARCHENCEPHALA ARCHONTIA.)

(PRIMATE SERIES.)

ORDER I.—PRIMATES.

SUB-ORDER ANTHROPOIDEA.

(Bimana.)

1. Hominidae   = Anthropini, Huxl., M. T. & G., 1864, i, 153.

(Simiae.)

(Simiae catarrhinae.)

2. Simiidae   = Anthropomorpha, Huxl., M. T. & G. 1864, i, 648.

a. Simiinae === Simiina, Gray, M., L., &
Fr.-cat. B., 6.

b. Hylobatinae === Hylobatina, Gray, M., L.,
& Fr.-cat. B., 9.

3. Cynopithecidae === Cynopithecini, Huxl., M. T.
& G., 1864, i, 671.

a. Semnopithecinae === Sub-Family II, Mart., Man
and Monkeys, 415.

b. Cynopithecinae === Sub-Family III, Mart., Man
and Monkeys, 503.

(*Simiae platyrhinae.*)

4. Cebidae === Platyrhini, Huxl., M. T. & G.,
1864, ii, 93.

a. Mycetinae === Mycetinae, Miv., P. Z. S.,
1865, 547.

b. Cebinae === Cebinae, Miv., P. Z. S.,
1865, 547.

c. Nyctipithecinae === Nyctipithecinae, Miv., P.
Z. S., 1865, 547.

d. Pitheciinae === Pitheciinae, Miv., P. Z. S.,
1865, 547.

5. Mididae === Arctopithecini, Huxl., M. T.
& G., 1864, ii, 124.

SUB-ORDER PROSIMIAE.

(*Lemuroidea.*)

6. Lemuridae === Lemuridae, Geoff., Cat. Primates, 66.

a. Indrisinae	= Indrisinae, Miv., P. Z. S., 1866, 151.
b. Lemurinae	= Lemurinae, Miv., P. Z. S. 1867, 960.
c. Nycticebinae	= Nycticebinae, Miv., P. Z. S., 1864, 643.
d. Galagininae	= Galagininae, Miv., P. Z. S., 1864, 645.
7. Tarsiidae	= Tarsidae, Geoff., Cat. Primates, 83.

(*Daubentonioidea.*)

8. Daubentoniidae	= Cheiromyidae, Geoff., Cat. Primates, 85.

(Feral Series.)

Order II.—FERÆ.

Sub-Order Fissipedia.

(*Aeluroidea.*)

(*Aeluroidea typica.*)

9. Felidae	= Felidae, Fl., P. Z. S., 1869, 15–18.
a. Felinae	= Felidae, § 1, Gray, P. Z. S., 1867, 261.
b. Guepardinae	= Felidae, § 2, Gray, P. Z. S., 1867, 277.

c. *Machaerodontinae* > Felinae, Burm., A. M. P. B.
-A. i, 122–138.

10. Cryptoproctidae = Cryptoproctidae, Fl., P. Z. S., 1869, 22.

(*Aeluroidea hyaeniformia.*)

11. Protelidae = Protelidae, Fl., P. Z. S., 1869, 27, 474.

12. Hyaenidae = Hyaenidae, Fl., P. Z. S., 1869, 26.

(*Aeluroidea viverriformia.*)

13. Viverridae = Viverridae, Fl., P. Z. S., 1869, 18.

a. Viverrinae { Viverrina, } Gray, C. P. &
{ Genettina, } E. M., 46, 49.

b. Prionodontinae = Prionodontina, Gray, C. P. & E. M., 52.

c. Galidiinae = Galidiina, Gray, C. P. & E. M., 55.

d. Hemigalinae = Hemigalina, Gray, C. P. & E. M., 56.

e. Arctictidinae = Arctictidina, Gray, C. P. & E. M., 57.

f. Paradoxurinae = Paradoxurina, Gray, C. P. & E. M., 59.

g. Cynogalinae = Cynogalina, Gray, P. Z. S., 1867, 521. = *Cynogalidae.*

h. Herpestinae = Herpestina, Gray, C. P. & E.
M., 144. (h-i<*Herpestidae.*)

i. Cynictidinae = Cynictidina, Gray, C. P.
& E. M., 169.

j. Rhinogalinae = Rhinogalina, Gray, C.P.&E.
M., 172. j-k<*Rhinogalidae.*

k. Crossarchinae < Crossarchina, Gray, C. P.
& E. M., 176.

14. Eupleridae = Eupleréens, Doy., A. S. N.,
2e s., iv, 1835, Z., 281.

(*Cynoidea.*)

15. Canidae = Canidae, Fl., P. Z. S., 1869,
23.

a. Caninae = Canidae, Gray, C. P. & E. M.,
178.

b. Megalotinae = Megalotidae, Gray, C. P. &
E. M., 210.

(*Arctoidea.*)

(*Arctoidea musteliformia.*)

16. Mustelidae = Mustelidae, Fl., P. Z. S., 1869,
11–14.

a. Mustelinae = Mustelina, Gray, C. P. &
E. M., 81.

b. Melinae = Melina, Gray, C. P. & E. M.,
122. (b-f < *Melinidae.*)

c. Mellivorinae = Mellivorina, Gray, C. P. & E. M., 131.

d. Mephitinae = Mephitina, Gray, C. P. & E. M., 133.

e. Zorillinae = Zorillina, Gray, C. P. & E. M., 139.

f. Helictidinae = Helictidina, Gray, C. P. & E. M., 141.

g. Lutrinae = Lutrina, Gray, C. P. & E. M. 100. (g-h < *Mustelidae*.)

h. Enhydrinae = Enhydrina, Gray, C. P. & E. M., 118.

(*Arctoidea typica.*)

17. Ursidae = Ursidae, Fl., P. Z. S., 1869, 6–9.

(*Arctoidea procyoniformia.*)

18. Acluridae = Ailuridae, Fl., P. Z. S., 1869, 11, 36.

19. Cercoleptidae > Procyonidae, Fl., P. Z. S., 1869, 9, 32.

20. Procyonidae > Procyonidae, Fl., P. Z. S., 1869, 9, 32.

a. Nasuinae = Nasuidae, Gray, C. P. & E. M., 238.

b. Procyoninae = Procyonidae, Gray. C. P. & E. M., 242.

21. Bassarididae   = Bassaridae, Gray, C. P. & E. M., 245.

(*Fissipedia sedis incertae.*)

22. *Simocyonidae*   = *Famille aujourd'hui éteinte*, Gaudry, (320), 37.

23. *Arctocyonidae*,   < *Arctocyoninae*, Giebel, Säugethiere, 755.

?

24. *Hyaenodontidae*   = *Hyaenodontidae*, Leidy, Ext. Mamm. Dak. & Neb., 38.

(𝕓   𝕢)

SUB-ORDER PINNIPEDIA.

(*Phocoidea.*)

25. Otariidae   = Otariadae, Allen, B. M. C. Z., ii, 1; Gill, A. N., iv., 675.

26. Phocidae   = Phocidae, Gill, C. E. I., 1866, 5, 8.

   a. Phocinae   = Phocinae, Gill, C. E. I., 1866, 5.

   b. Cystophorinae   = Cystophorinae, Gill, C. E. I., 1866, 6.

   c. Stenorhynchinae   = Stenorhynchinae, Gill, C. E. I., 1866, 6.

(*Rosmaroidea.*)

27. Rosmaridae   = Rosmaridae, Gill, C. E. I., 1866, 7.

ORDER III.—UNGULATA.

SUB-ORDER ARTIODACTYLI.

(*Pecora s. Ruminantia.*)

(*Pecora ? edentata.*)

27a. *Chalicotheriidae*   = *Chalicotherium*, Falc., Pal. Mem., i, 190, 208, 523.

(*Pecora tylopoda s. phalangigrada.*)

28. Camelidae   = Camélidés, Gerv., Mamm. ii. 223.

(*Pecora unguligrada.*)

(*Pecora unguligrada typica.*)

(*Girafoidea.*)

29. Giraffidae   = Girafidés, Gerv. Mamm. ii, 210.

(*Booidea.*)

(*Booidea typica.*)

30. Saigiidae   = Saigiinae, Mur., P. Z. S., 1870, 451.

31. Bovidae   = Bovidés, Gerv., Mamm. ii, 174.

    a. Bovinae   = Bovina, Rutim., N. D. S. G. N., xxiii, 21.

b. Ovibovinae	< Boveae, Gray, Mamm., iii, 15.
c. Antilopinae	{ Antilopeae, } G. M., iii, { Strepsicereae, } 45, 131.
d. Caprinae	= Capreae, Gray, Mamm., iii, 142.
e. Ovinae	= Oveae, Gray, Mamm., iii, 160.
32. Antilocapridae	= Antilocapridae, Mur., P. Z. S., 1870, 334.

(*Booidea cerviformia.*)

33. Cervidae	= Cervidae, Scl., P. Z. S., 1870, 114.
a. Cervinae	= Cervinae, Scl., P. Z. S., 1870, 114.
b. Cervulinae	= Cervulinae, Scl., P. Z. S., 1870, 115.
c. Moschinae	= Moschinae, Scl., P. Z. S., 1870, 115.

(*Pecora unguligrada traguloidea.*)

34. Tragulidae	= Tragulidae, A. Milne Edw., A. S. N.., 5e s., ii, Z., 1864, 157.

(*Pecora unguligrada incertae sedis.*)

35. *Sivatheriidae*	= *Sivatherium*, Falc., Pal. Mem., i, 247.

36. *Helladotheriidae* = *Famille aujourd'hui éteinte,*
Gaudry, A. F. Att. (321), 252.

(*Pecora dentata.*)

(*Oreodontoidea.*)

37. *Oreodontidae.*

a. *Oreodontinae* = *Oreodontidae,* Leidy, Ext.
Mamm. Dak. & Neb., 71.

b. *Agriochoerinae* = *Agriochoeridae,* Leidy, Ext.
Mamm. Dak. & Neb., 131.

(*Anoplotheroidea.*)

38. *Anoplotheriidae* = *Anoplotheriidae,* Leidy, Ext.
Mamm. Dak. & Neb., 206.

39. *Dichobunidae* = Moschidae § *Dichobunina,*
Turn, P. Z. S., 1849, 158.

(*Omnivora.*)

(*Merycopotamoidea.*)

40. *Merycopotamidae* = *Merycopotamus,* Falc., Pal.
Mem., ii. 407.

(*Hippopomatoidea.*)

41. Hippopotamidae = Hippopotamidae, Gray, C.
P. & E. M., 356.

a. Hippopotaminae = Hippopotamus, Falc., Pal.
Mem., i, 130.

b. Choeropsinae = Choeropsis, A. Milne Ed.,
R. H. N. M., 43.

(*Setifera suiformia.*)

42. Phacochoeridae = Phacochoeridae, Gray, B. M., 352.

43. Suidae = Suidae, Gray, C. P. & E. M., 327.

(*Setifera dicotyliformia.*)

44. Dicotylidae = Dicotylidae, Gray, C. P. & E. M., 350.

(*Anthracotheroidea.*)

45. *Anthracotheriidae* < Hippopotamidae, Turn., P. Z. S., 1849, 157.

   a. *Hyopotaminae* < *Anthracotheriidae*, L'dy, Ex. Mamm. Dak. & Neb., 202.

   b. *Anthracotheri-* < *Anthracotheriidae*, L'dy, Ex. *inae* Mamm. Dak. & Neb., 202.

SUB-ORDER PERISSODACTYLI.

(*Anchippodontoidea.*)

45a. *Anchippodontidae* = Trogosus, Leidy, P. A. N. S., Phil., 1871, 114.

(*Solidungula.*)

46. Equidae = Equidae, Gray, C. P. & E. M., 262.

47. *Anchitheriidae*   = *Anchitheridae*,   Leidy,   Ext.
Mamm. Dak. & Neb., 302.

(*Multungula.*)

(*Rhinocerotoidea.*)

(*Rhinocerotoidea rhinocerotiformia.*)

48. Rhinocerotidae   = Rhinocerotidae, Gray, C. P. &
E. M., 295.

(*Rhinocerotoidea macraucheniiformia.*)

49. *Macraucheniidae*   = *Macrauchenia*, Burm., A. M.
B.-A., i, 32, 1864.
50. *Palaeotheriidae*   < *Palaeothérioïdes*, Pictet, Paléont., 2e ed., i, 309–313.

(*Tapiroidea.*)

51. Tapiridae   = Tapiridae, Gray, C. P. & E.
M., 252.
52. *Lophiodontidae*   < Tapiroïdes, Pictet, Paléont.,
2e ed., i, 301.

(*Pliolophoidea.*)

53. *Pliolophidae*   = *Pliolophus*,   Owen,   Pal.,
1860, 325.

(*Perissodactyli? incertae sedis.*)

54. *Elasmotheriidae*   Rhinocéroides, Pictet, Paléont., 2e ed., i, 294.

ORDER IV.—TOXODONTIA.

55. *Nesodontidae*   = *Nesodon*, Owen, Ph. T.,
       1853, 291.

56. *Toxodontidae*   = *Toxodon*,. Burm, A. M.
       B.-A., i, 254, 1864.

ORDER V.—HYRACOIDEA.

57. Hyracidae   = Hyracidae, Gray, C. P. & E.
       M., 279.

ORDER VI.—PROBOSCIDEA.

58. Elephantidae   < Proboscideae, Falc., Pal.
       Mem., ii, 1868.

   Elephantinae   = Elephantidae, Gray, C. P. &
       E. M., 358.

   Mastodontinae   = *Mastodontidae*, Gray, C. P. &
       E. M., 359.

59. *Dinotheriidae*   = [*Dinothériides*,] Gaudry, An.
       F. Att., 321, 162.

MUTILATE SERIES.

ORDER VII.—SIRENIA.

(*Halicoroidea*.)

60. *Halitheriidae*   < Halicorida, Brandt, Symb.
       Siren., ii, (f. 3,) 344.

61. Halicoridae   < Halicorida, Brandt, Symb.
                     Siren., ii, (f. 3,) 344.

62. Rhytinidae   < Halicorida, Brandt, Symb.
                     Siren., ii, (f. 3,) 344.

(*Manatoidea.*)

63. Trichechidae   = Manatida, Brandt, Symb.
                     Siren., ii, (f. 3,) 343.

Order VIII.—CETE.

Sub-Order Zeuglodontes.

64. *Basilosauridae*   < *Zeuglodontes*, VanBen., Mém.
                     Ac. R. Belg., xxxv, 1865.

65. *Cynorcidae*   = *Cynorcidae*, Cope, P. A. N. S.,
                     1867, 144.

Sub-Order Denticete.

(*Delphinoidea.*)

(*Delphinoidea platanistiformia.*)

66. Platanistidae   < Platanistidae, Fl., Trans. Zool.
                     Soc., vi, 113, 1867.

67. Iniidae   < Platanistidae, Fl., Trans.
                     Zool. Soc., vi, 114, 1867.

(*Delphinoidea typica.*)

68. Delphinidae   > Delphinidae, Fl., Trans. Zool.
                     Soc., vi, 113, 1867.

a. Pontoporiinae   = Pontoporiinae, Gill, C. E. I., vi., 124, 1871.

b. Delphinapterinae = Beluginae, Fl., Trans. Zool. Soc., vi, 115, 1867.

c. Delphininae   < Delphininae, Fl., Trans. Zool. Soc., vi, 115, 1867.

d. Globiocephalinae < Delphininae, Fl., Trans. Zool. Soc., vi, 115, 1867.

(*Delphinoidea ziphiiformia.*)

69. Ziphiidae   = Ziphioïdes, Fisch, N. A. M. H. N. P., iii, 41, 1867.

a. Ziphiinae   = Ziphiinae, Gill, C. E. I., vi, 124, 1871.

b. Anarnacinae   = Anarnacinae, Gill, C. E. I., vi, 124, 1871.

(*Physeteroidea.*)

70. Physeteridae   = Physeteridae, Gill, A. N., iv, 727, 1871.

a. Physeterinae   = Physeterinae, Gill, A. N., iv, 732, 1871.

b. Kogiinae   = Kogiinae, Gill, A. N., iv, 732, 1871.

(*Denticete incertae sedis.*)

71. *Rhabdosteidae*   = *Rhabdosteidae*, Gill, C. E. I., vi, 123, 1871.

SUB-ORDER MYSTICETE.

72. Balaenopteridae = Balaenopteridae, Fl., Proc. Zool. Soc., 1864, 291.

   a. Agaphelinae = Agaphelinae, Gill, C. E. I., vi, 124, 1871.

   b. Megapterinae = Megapterinae, Fl., Proc. Zool. Soc., 1864, 391.

   c. Balaenopterinae = Balaenopterinae, Fl., Proc. Zool. Soc., 1864, 391.

73. Balaenidae = Balaenidae, Fl., Proc. Zool. Soc., 1864, 389.

SUPER-ORDER INEDUCABILIA.

(LISSENCEPHALA Owen=MICROSTHENA Dana.)

(INSECTIVOROUS SERIES.)

ORDER IV.—CHIROPTERA.

SUB-ORDER ANIMALIVORA.

(*Hæmatophilina.*)

74. Desmodidae = Haematophilini, Huxl., P. Z. S. L., 1865, 386.

(*Histiophora.*)

75. Phyllostomidae > Phyllostomidae, Gray, P. Z. S. L., 1866, 111.

76. Mormopidae = Mormopes, Car., Handb. Zool., i, 83.

77. Rhinolophidae   < Rhinolophidae, Gray, P. Z.
S. L., 1866, 81.

78. Megadermidae   < Megadermata, Pet., M. P. A.
W. Berlin, 1865, 256.

a. Vampyrinae   = Vampiri, Pet., M. P. A. W.
Berlin, 1865, 503.

b. Glossophaginae   = Glossophagae, Pet., M. P.
A. W. Berlin, 1868, 361.

c. Stenoderminae   = Stenodermata, Pet., M. P. A.
W. Berlin, 1865, 356, 524.

(*Gymnorhina*.)

79. Vespertilionidae   = Vespertiliones, Pet., M. P. A.
W. Berlin,1865,258,524.

a. Vespertilioninae   = Vespertilioniens,Gerv., An.
Am.S.Cast.—Mamm.,74.

b. Nycticejinae   = Nycticéins, Gerv., Mamm.,
74.

80. Molossidae   = Molossi, Pet., M. P. A. W.
Berlin, 1865, 573.

81. Noctilionidae   = Brachyura, Pet., M. P. A. W.
Berlin, 1865, 257.

a. Noctilioninae   = Noctilionins,Gerv., An.Am.
S. Cast.—Mamm., 52.

b. Emballonurinae   = Noctilionins,Gerv.,An.Am.
S. Cast.—Mamm., 62.

c. Furiinae = Furia, Gerv., An. Am. S. Cast.—Mamm., 69.

Sub-Order Frugivora.

82. Pteropodidae = Pteropi, Pet., M. P. A. W. Berlin, 1867, 320, 867.

Order VI.—INSECTIVORA.

Sub-Order Dermoptera.

83. Galeopithecidae = Galeopithecidae, Miv., J. A. & P., ii, 1868, 124.

Sub-Order Insectivora Vera.

(Soricoidea.)

84. Talpidae = Talpidae, Miv., J. A. & P., ii, 1868, 150.

  a. Talpinae = Talpina, Miv., J. A. & P., ii, 1868, 151,

  b. Myogalinae = Myogalina, Miv., J. A. & P., ii, 1868, 152.

85. Soricidae = Soricidae, Miv., J. A. & P., ii, 1868, 153.

(Erinaceoidea.)

86. Erinaceidae = Erinaceidae, Miv., J. A. & P., ii, 1868, 146.

  a. Erinaceinae = Hérissons, Gerv., H. N. Mamm., i, 229.

b. Gymnurinae   = Gymnures, Gerv., H. N. Mamm., i, 231.

(*Centetoidea.*)

87. Centetidae   = Centetidae, Miv., J. A. & P., ii, 1868, 147.

a. Centetinae   = Tanrecs, Gerv., H. N. Mamm., i, 233.

b. Solenodontinae   = Solénodontes, Gerv., H. N. Mamm, i, 246.

88. Potamogalidae   = Potamogalidae, Allm., T. Z. S., vi, 149, 1–16.

(*Chryschloridoidea.*)

89. Chrysochlorididae= Chrysochloridae, Miv., J. A. & P., ii, 1868, 150.

(*Macroscelidoidea.*)

90. Macroscelididae   = Macroscelididae, Miv., J. A. & P., ii, 1868, 143.

a. Rhynchocyoninae = Rhynchocyons, Gerv., H. N. Mamm., i, 238.

b. Macroscelidinae   = Macroscélidiens, Gerv., H. N. Mamm., i, 235.

91. Tupayidae   = Tupaiidae, Miv., J. A. & P., ii, 1868, 145.

(*Insectivora incertae sedis.*)

92. *Leptictidae*   < *Leptictis*, Leidy, Ext. Mamm. Dak. & Neb., 345.

(Rodent Series.)

Order GLIRES.

Sub-Order Simplicidentati.

(*Lophiomyoidea.*)

93. Lophiomyidae = Lophiomides, A. M. Edw., N.
A. M. H. N. P., iii, 114.

(*Myoidea.*)

94. Pedetidae = Pedetina, Car., Handb. Zool.,
i, 101.

95. Dipodidae = Dipodina, Car., Handb. Zool.,
i, 101.

96. Jaculidae = Jaculina, Car., Handb. Zool.,
i, 101.

97. Muridae = Muridés, Gerv., H. N. Mamm.,
i, 417.

a. Spalacinae = Rhizodontes a. Spalacini,
Br't., S. R., 307.

b. Georhychinae = Rhizodontes b. Georhy-
chini, Br't., S. R., 308.

c. Murinae = Murini, Lillj., Gnag.
Däggdj., 12.

d. Siphneinae = Prismatodontes b. Macro-
nyches, Br't., S. R., 309.

e. Ellobiinae = Primatodontes a. Brachyo-
nyches, Br't., S. R., 309.

f. Arvicolinae = Arvicolini, Lillj., Gnag. Däggdj., 22.

(*Myoxoidea.*)

98. Myoxidae = Myoxidae, Lillj., Gnag. Däggdj., 31.

(*Saccomyoidea.*)

99. Saccomyidae = Saccomyinae, Bd., M. N. A., 405. (e Saccomyidiis.)

100. Geomyidae = Sciurospalacoïdes, Br't., S. R., 301.

(*Castoroidea.*)

101. Castoridae = Castoridae, Morgan, Am. Beaver, 186.

(*Sciuroidea.*)

102. Sciuridae = Sciurida, Car., Handb. Zool., i, 96.

a. Sciurinae = Campsiurina, Car., Handb. Zool., i, 96.

b. Arctomyinae = Arctomyina, Car., Handb. Zool., i, 97.

(*Anomaluroidea.*)

103. Anomaluridae = Anomalurina, Car., Handb. Zool., i, 98.

(Haploodontoidea.)

104. Haploodontidae = Haploodontidae, Lillj., Gnag.
Däggdj., 41.

(Hystricoidea.)

105. Spalacopodidae = Spalacopodidae, Lillj., Gnag.
Däggdj., 44.

 a. Octodontinae   × Octodontina, Waterh., N.
         H. Mamm., ii, 242.

 b. Ctenodactylinae   < Octodontina, Waterh., N.
         H. Mamm., ii, 242.

 c. Echimyinae   < Echimyina, Waterh., N. H.
         Mamm., ii, 286.

 d. Cercolabinae   = Cercolabina, Waterh., N.
         H. Mamm., ii, 484, (398).

106. Hystricidae   < Hystrichina, Car., Handb.
         Zool., i, 109.

107. Dasyproctidae   = Dasyproctina, Car., Handb.
         Zool., i, 110.

 a. Dasyproctinae   = Dasyproctiens, Gerv., H. N.
         Mamm., 327.

 b. Coelogenyinae   = Célogényens, Gerv., H. N.
         Mamm., 325.

108. Caviidae   < Caviina, Car., Handb. Zool.,
         i, 110.

109. Hydrochoeridae < Caviina, Car., Handb. Zool.,
         i, 110.

110. Chinchillidae = Chinchillidae, Lillj., Gnag. Daggdj., 42.

a. Chinchillinae = Orobii *seu* Eriomyes monti-colae, Br't., S. R., 317.

b. Lagostominae = Homalobii *seu* Eriomyes planicolae,Br't.,S.R.,317.

SUB-ORDER DUPLICIDENTATI.

111. Lagomyidae = Lagomyidae, Gray, A. & M. N. H., xx, 219, 1867.

112. Leporidae = Leporidae, Gray, A. & M. N. H., xx, 219, 1867.

ORDER XIII.—BRUTA.

SUB-ORDER VERMILINGUIA.

113. Myrmecophagi- = Myrmecophagidae, Gray, C.
dae   P. & E. M., 390.

a. Myrmecophaginae {Myrmecophaga, } Gray, C. P.
                   {Tamandua,   } &E.M.,390.

b. Cyclothurinae = Cyclothurus, Gray, C. P. & E. M., 392.

SUB-ORDER SQUAMATA.

114. Manididae = Manididae, Gray, C. P. & E. M., 366.

SUB-ORDER FODIENTIA.

115. Orycteropodidae = Orycteropodidae, Gray, C. P. & E. M., 389.

SUB-ORDER TARDIGRADA.

116. Bradypodidae   = Bradypodidae, Gray, C. P. &
                            E. M., 362

    a. Bradypodinae   { Bradypus,   } Gray, 363,
                     { Arctopithecus, } 364.

    b. Choloepodinae   = Choloepus, Gray, C. P. &
                         E. M., 363.

117. *Megatheriidae*   = *Gravigrada*, Burm., A. M. P.
                         B. A., i, 32.

    a. *Megatheriinae*

    b. *Mylodontinae*

SUB-ORDER LORICATA.

118. Dasypodidae   > Dasypodidae, Gray, P. Z. S.,
                      1865, 360.

    a. Dasypodinae   < Dasypodina, Gray, P. Z. S.,
                     1865, 360.

    b. Tatusiinae   < Dasypodina, Gray, P. Z. S.,
                     1865, 360.

    c. Xenurinae   < Dasypodina, Gray, P. Z. S.,
                     1865, 365.

    d. Tolypeutinae   = Tolypeutina, Gray, P. Z. S.,
                     1865, 365.

119. Chlamydophori- = Chlamyphoridae, Gray, P. Z.
      dae   S., 1865, 387.

120. *Hoplophoridae*   = *Hoplophoridae*, Huxl., Phil.
                         Trans., clv, 31.

?

121. *Ancylotheriidae* = *Famille aujourd'hui eteinte*, Gaudry, An. foss. d'Att., i, 129, 321.

SUB-CLASS DIDELPHIA.

ORDER XIV.—MARSUPIALIA.

SUB-ORDER RHIZOPHAGA.

122. Phascolomyidae = Phascolomyidae, Waterh., N. H. Mamm., i, 241.

SUB-ORDER SYNDACTYLI.

(*Poephaga.*)

123. Macropodidae = Macropodidae, Waterh., N. H. Mamm., i, 50.

(*Carpophaga.*)

124. Tarsipedidae = Tarsipédidés, Gerv., Mamm., ii, 277.

125. Phalangistidae = Phalangistidae, Owen, T. Z. S., ii, 332.

   a. Petaurinae = Petauristins, Gerv., H. N., Mamm., ii, 276.

   b. Phalangistinae = Phalangistins, Gerv., H. N. Mamm., ii, 274.

126. Phascolarctidae = Phascolarctidae, Owen, T. Z. S., ii, 332.

(*Diprotodontoidea.*)

127. *Diprotodontidae*   < Diprotodon, Owen, Palæont., 394–395.

128. *Thylacoleonidae*   < Thylacoleo, Fl., Jour. Geol. S. L., xxiv, 1868, 307.

(*Entomophaga.*)

129. Peramelidae   = Peramelidae, Waterh., N. H. Mamm., i, 354.

   a. Chœropodinae

   b. Peramelinae

SUB-ORDER DASYUROMORPHIA.

130. Dasyuridae   = Dasyuridae, Owen, T. Z. S., ii, 332.

   a. Sarcophilinae

   b. Dasyurinae

   c. Phascogalinae

131. Myrmecobiidae   = Ambulatoria, Owen, T. Z. S., ii, 332.

SUB-ORDER DIDELPHIMORPHIA.

132. Didelphididae   = Didelphididae, Waterh., N. H. Mamm., ii, 462.

MARSUPIALIA INCERTAE SEDIS.

133. *Plagiaulacidae*   = Plagiaulax, Falc., Journ.
Geol. S. L., 1862, 348.

134. *Dromatheriidae*   = Dromatherium, Owen, Pal.,
302.

SUB-CLASS ORNITHODELPHIA.

ORDER XVI.—MONOTREMATA.

SUB-ORDER TACHYGLOSSA.

135. Tachyglossidae > Ornithorhynchidae, Gray, C.
P. & E. M., 393.

SUB-ORDER PLATYPODA.

136. Ornithorhynchi- > Ornithorhynchidae, Gray, C.
dae   P. & E. M., 393.

.

BIBLIOGRAPHY,

OR

OF AUTHORS REFERRED TO

LIST OF AUTHORS REFERRED TO.

The following enumeration of works is chiefly intended to explain the abbreviations used in connection with the preceding list of families: the works most accessible to students generally have been used, whenever they could be referred to in explanation of the limits of families adopted; special monographs have been chiefly referred to when the groups in connection with which they are cited have not been limited in the same manner in general works. The "Ostéographie" of de Blainville, although not actually referred to in connection with any special family, is so indispensable to any investigator of the mammals, and has been so much used by the writer, that the title thereof and an analysis of its contents have been given; the analysis and assignment of dates of publication of the several monographs will doubtless prove useful, and save to some time and labor like that necessarily devolved upon the writer in ascertaining the data furnished.

For the information of students, and because it is information often desired, the publishers' prices of most of the works cited are given, in the currency of the country where they were published. Many of the separate monographs reprinted from journals can be obtained from the second-hand book dealers—especially the German—and from the Naturalists' Agency of Salem, Mass., but at varying prices.

In order to secure uniformity of typography, only the initial letters of the characteristic words are capital, the example of the learned brothers Grimm, as well as other German writers, sanctioning such usage for their language. The initial letters, however, of the more important words of the general titles, and to which reference is made in the list, are capitalized, corresponding with and rendering at once intelligible the abbreviated references. The punctuation of the respective title-pages is adopted. The symbol (<) denotes that the memoir after which it is inserted is contained in the volume or series whose title follows; the symbol of equality (=) denotes that the memoir is co-extensive with the volume.

ALLEN (Joel Asaph). On the eared seals (Otariadæ), with detailed descriptions of the North Pacific species, by J. A. Allen. Together with an account of the habits of the northern fur seal (Callorhinus ursinus), by Charles Bryant. < Bulletin of the Museum of Comparative Zoology, · · · . II, No. 1 = pp. 1—108.

ALLMAN (George James). On the characters and affinities of Potamogale. · · · . < Transactions of the Zoological Society of London, VI, 1—16, pl. 1-2, 1866.

BAIRD (Spencer Fullerton). Mammals of North America; the descriptions of species based chiefly on the collections in the museum of the Smithsonian institution. · · · . With eighty-seven plates of original figures, illustrating the genera and species, and including details of external form and osteology.

(31)

Philadelphia: J. B. Lippincott & Co., 1859. [4to., 4 p. l., xi—xxxiv, 735 pp. + (Part II, 1—55 pp.) 736—764 pp., 87 pl. (29 col.)—$10; with col. pl., $15.]

[" Part I. General report upon the Mammals of the several Pacific railroad routes. .... Washington, D. C., July, 1857:" reprinted from the "Reports of explorations and surveys to ascertain the most practicable and economical route for a railroad from the Mississippi river to the Pacific Ocean. .... Volume VIII. Washington : .... 1857." (60 pl. in v. VI, VII, VIII.) "Part II. Special report upon the Mammals of the Mexican boundary. By Spencer F. Baird, ... With notes by the naturalists of the survey. Washington, D. C., January, 1859:" reprinted from the "Report on the United States and Mexican boundary survey, made under direction of the secretary of the interior, by William H. Emory, major first cavalry and United States commissioner. Volume II. Washington : .... 1859. (Part II. [§1.] Mammals of the boundary, ...)" 62 pp. 27 pl.]

BLAINVILLE (Henri Marie Ducrotay de). Ostéographie ou description iconographique comparée du squelette et du système dentaire des Mammifères recents et fossiles pour servir de base à la zoologie et à la géologie | par H. M. Ducrotay de Blainville .... Ouvrage accompagné de 323 planches lithographiées sous sa direction par M. J. C. Werner, peintre du Museum d'histoire naturelle de Paris, précédé d'une étude sur la vie et les travaux de M. de Blainville, par M. P. Picard.—[I—IV.—See "Contents."]—Paris : J. B. Baillière et fils .... 1839-1864. [Text, 4to., 4v.; Atlas, fol., 4v.]

[Published in twenty-six fascicules ; the first twenty-five under the title : "Ostéographie ; ou, description iconographique comparée du squelette et du système dentaire des cinq classes d'animaux vertébrés récents et fossiles, pour servir de base à la zoologie et à la géologie par M. H. M. Ducrotay de Blainville .... Ouvrage accompagné de planches lithographiées sous sa direction par M. J. C. Werner .... Paris, Arthus Bertrand, ..." [1839-1855.] The twenty-sixth and last fascicule was issued with the *special* title above given, titles for the four volumes of text and four of plates, table of contents and index, by the Baillières in 1864. The subscription price was 2 francs 35 centimes per plate ; the price of the twenty-sixth livraison, 45 francs ; and of the whole, on completion, 800 francs, "au lieu de 961 fr."

The culpable neglect of the publishers to give the dates of publication of the several fascicules has doubtless devolved upon many investigators, as upon the writer, much trouble and annoyance in ascertaining them, and to save to others similar trouble, a collation is here presented, the dates having chiefly been ascertained from Wagner's annual reports in the "Archiv für Naturgeschichte." The appearance of successive fascicules has not been noticed in the "Bibliographie de la France."

The titles of the respective monographs given below are those at the upper fourth of the first page of each monograph, and which are the only special titles published.

The work is more remarkable as a methodical repertory of facts respecting superficial osteological details, than as a digest exhibiting acute appreciation of the value and subordination of characters and their taxonomical application, or orthodox views respecting classification and the geological succession of animals—the concurrent views of the most recent and approved investigators being the standard. The "genera," it must be remembered, are generally about equal in extent to the families now generally adopted.]

CONTENTS.

Tome premier | Primatès—Secundatès | Avec atlas de 59 planches. [7 pp + 9 parts*, as below:—]
Atlas—Tome premier | composé de 59 planches | Primatès—Secundatès. [3 p. l. + 5 parts, viz.:—]

[A title-page with the more *general* title [see above] and the addition:—"Mammifères—Tome premier" was issued with the first fascicule in "1839," and another with the modification "Mammifères.—Primatès : Pithecus. Cebus. Lemur." in "1841," but both are superseded by the *special* title issued for the first volume with the twenty-sixth fascicule.]

(Etude sur la vie et les travaux de M. de Blainville, par M. P. Nicard.) [1864.—ccxxiii. pp. < F. xxvi.]
([*A.*] De l'ostéographie en général. > Ostéographie des Mammifères. pp. 19–47). [1839.—47 pp. < F. i.]
([*B.*] Ostéographie des Primatès.—Sur les primatès en général et sur les singes (*Pithecus*) en particuliér.) [1839.—52 pp. 11 pl. < F. i. (+ pl. 1 *bis* and 5 *bis*. < F. xxv, 1855.)]

[A *secondary general* title for the Primatès was issued as the first pages (p 1 = 1 1) of the preceding, viz.: " Ostéographie des Mammifères de l'ordre des Primatès, suivie de recherches sur l'histoire de la science à leur égard, les principes de leur classification, leur distribution géographique actuelle et leur ancienneté à la surface de la terre."]

([*C.*] Ostéographie des Primatès.—Sapajous (*Cebus*).) [1839.—31 pp. 9 pl. = F. ii.]
([*D.*] Ostéographie des Primatès.—Makis (*Lemur*).) [1839.—48 pp. 11 pl. < F. iii.]
([*E.*] Mémoire sur la véritable place de l'Aye-Aye dans la série des Mammifères. Lu à la Société philomatique, le 16 mai 1816.) [1839.—40 pp. < F. iii.—Plate < F. iii.—Plate =pl. 5 < D.]
([*F.*] De l'ancienneté des Primatès à la surface de la terre.) 68 pp. [1839] < F. iv.—Saus planches.]
([*G.*] Ostéographie des Cheiroptères (*Vespertilio*, L.).) [1839.—104 pp. 15 pl. < F. v.]
([*H.*] Ostéographie des Mammifères insectivores (*Talpa, Sorex* et *Erinaceus*, L.)) [1840.—113 pp. 11 pl. = F. vi.]

Tome deuxième | Secundatès | Avec atlas de 117 planches. [viii. pp. + 9 parts.]
Atlas—Tome deuxième | composé de 117 planches | Secundatès. [2 p. l. + 8 parts, viz.:—]

([*I.*] Ostéographie des Carnassiers. [1840.—85 pp. < F. vii.]

[A *secondary* title for the Carnassiers (I—Q) was issued as the first pages (p. 1 = 1 1) of the preceding, viz.: " Ostéographie des Carnassiers, précédée de considérations sur l'histoire de la science à leur égard, les principes de leur classification, leur distribution géographique actuelle, et suivie de recherches sur leur ancienneté à la surface de la terre."]

([*J.*] Des Phoques (G. *Phoca*, L.).) [1840.—51 pp. 10 pl. < F. vii.]
([*K.*] Des Ours (G. *Ursus*).) [1841.—94 pp. 18 pl. = F. viii.]
([*L.*] Des Petit-ours (G. *Subursus*).) [1841.—123 pp. 16 pl. = F. ix. (+ pl. 17 < F. x, 1842.]

---

* The "parts" is each monograph or series distinguished by a special and complete pagination or numeration of plates.

[With this fascicule was issued a *general* title limited thus : "Mammifères.—Carnassiers : | Vespertilio. Talpa. Sorex. Erinaccus. Phoca. Ursus. Subursus. · · · · . 1841."]

([*M.*] Des Mustelas (G. *Mustela*, L.).) [1842.—83 pp. 15 pl. = F. x.]
([*N.*] Des Viverras.) [1842.—100 pp. 13 pl. = F. xi.]
([*O.*] Des Felis.) [1843.—196 pp, 1 folded tab., 19 pl. = F. xii. (+ pl. 20 < F. xxv., 1855.)]
([*P.*] Des Canis.) [1843.—160 pp. 16 pl. = F. xiii.]
([*Q.*] Des Hyènes.) [1844.—84 pp. 8 pl. = F. xiv.]

Tome troisième | Quaternatès | Avec atlas de 54 planches. [viii pp. + 5 parts.]

Atlas—Tome troisième | composé de 54 planches | Quaternatès. [2 p. l. + 5 parts, viz.:]

([*R* or *S*] Des Eléphants.) [1845.—367 pp. 18 pl. = F. xvi.]
([*S* or *T*.] Du Dinotherium.) [1845.—64 pp. 3 pl. = F. xvii.]
([*T* or *U*.] Des Lamantins (Buffon), (*Manatus*, Scopoli), ou Gravigrades aquatiques.) [1844.—140 pp. 11 pl. = F. xv.]
([*V*.] Des Damans (Buffon), (*Hyrax*).) [1845.—47 pp. 3 pl. = F. xviii.]
([*V* or *X*.] Des Rhinocéros (Buffon), (G. *Rhinocéros*, L.).) [1846.—232 pp. 14 pl. = F. xx.]
([*X* and non-lettered.*] Monographie du Cheval. G. *Equus*.) 1864. [80 pp. < F. xxvi.]

Tome quatrième—Quaternatès—Maldentés | Avec atlas de 93 planches. [viii. pp. + 8 parts.]

Atlas—Tome quatrième | composé de 93 planches | Quaternatès—Maldentés. [2 p. l. + 11 parts.]

([*Y*.] Des Palæotheriums, Lophiodons, Anthracotheriums, Choeropotames.) [1846.—196 pp. 8 + 3 + 3 + 1 [= 15] pl. = F. xxi.]
([*Z*.] Des Tapirs (Buffon). (G. *Tapirus*, Brisson).) [1846.—52 pp. 6 pl. = F. xix.]
([*AA*.] Sur les Hippopotames (Buffon), (*Hippopotamus*, L.) et les Cochons (Buffon), (*Sus*, L.).) 1847. [248 pp. 8 + 9 [=17] pl. < F. xxii.]
([*BB*.] Des Anoplothériums (G. Cuvier) et sur les genres plus ou moins différents: 1849. [155 pp. 9 pl. = F. xxiii.]

Xiphodon,		Merycopotamus,	Falconer et Cauteley,†
Dichobune,	G. Cuvier, 1822.	Hippohyus,	1847.
Adapis,		Paloplotherium,	
Chalicothérium, J. Kaup, 1833.		Dichodon,	R. Owen, 1848.
Cainothérium, Brarard, 1835.		Hyopotamus,	
Microchoerus, &c.† Wood, 1846.)			

([*CC*.] Des Ruminants (*Pecora*, L.) en général et en particulier des Chameaux, des Lamas, Buffon. (G. *Camelus*, L.) 1850. [131 pp. 5 pl. = F. xxiv.]
([*DD*.] Ostéographie des Paresseux (*Bradypus*, L.).) [1840.—64 pp. 6 pl. = F. v.]
([*EE*. *General* title.] Publication posthume.—Explication des planches suivantes.
PILIFÈRES. Genres. Gorilla, Smilodon, Sciurus, Arctomys, Castor, Capromys, Myopotamus, Hystrix, Cavia, Equus, Cameloparda lis, Myrmecophaga, Macrotherium, Megatherium, Glyptodon, Toxodon, Elasmotherium, Macrauchenia et groupes qui s'y rattachent.
SQUAMMIFÈRES. Genre Crocodilus et groupes génériques voisins.
OSTÉOZOAIRES. Signification des os du crâne dans les diverses classes de ce type. 1855. [63 pp. 41 pl.]
Table alphabétique des quatre volumes. 1855. [lxvi. pp. < F. xxvi.]

BRANDT (Johann Friedrich). Symbolæ sirenologicæ, [fasciculus I,] quibus praecipue Rhytinæ historia naturalis illustratur. · · · · . (1845). < Mémoires

---

de l'Académie Impériale des Sciences de St. Pétersbourg. Sixième série. Sciences mathématiques, physiques et naturelles. Tome VII. Seconde partie: Sciences naturelles. Tome VII. . . . . 1849.—Zoologie et physiologie, 1—160, pl. 1—5.

BRANDT (Johann Friedrich). Beiträge zur nähern kenntniss der säugethiere Russland's. Von J. F. Brandt. (1851.) < Ib. Sixième série. Sciences mathématiques, physiques, et naturelles. Tome IX. Seconde partie. Sciences naturelles. Tome VII. . . . . 1855.—Zoologie et physiologie. 1—365. [Vierte abhandlung. Blicke auf die allmäligen fortschritte in der gruppirung der Nager [Glires] mit specieller beziehung auf die geschichte der gattuug Castor, besonders des altweltlichen Bibers. (pp. 77—124.) Fünfte abhandlung. Untersuchungen über die craniologischen entwickelungsstufen und die davon herzuleitenden verwandtschaften und classificationen der Nager der jetztwelt, mit besonderer beziehung auf die gattung Castor. (pp. 125—336, pl. i—xi + va.)]

—— Symbolæ sirenologicæ. Fasciculus II et III. Sireniorum, Pachydermatum, Zeuglodontum et Cetaceorum ordinis osteologia comparata, nec non Sireniorum generum monographiæ. . . . . . Petropoli, 1861—68. [4to., 3 p. l. 383 (+ 1) pp. 9 pl.] < Mémoires de l'Académie Impériale des Sciences de St. Pétersbourg, Sixième série. Sciences naturelles. 1—365, 19 pl.

—— De Dinotheriorum genere Elephantidorum familiæ adjugendo nec non de Elephaniidorum generum craniologia comparata. . . . . St. Pétersbourg, ... 1869. [4to. 1—38 pp.] < Ib. XIV, No. 1.

—— Untersuchungen über die gattung Klippschliefer (Hyrax Herm.), besonders in anatomischer und verwandtschaftlicher beziehung nebst bemerkungen über ihre verbreitung und lebensweise. . . . . St. Pétersbourg, 1869. . . . . [4to. vi, 127 pp. 3 pl.] < Ib. XIV, No. 2.

BURMEISTER (Carl Hermann Conrad, or, Hispanice, German). Descripcion de la Macrauchenia patachonica. < Anales del Museo público de Buenos Aires, . . . , para German Burmeister, director del Museo público de Buenos Aires. . . . . I, 32—65, pl. 1—4. 1864.

—— Fauna argentina.—Primera parte. Mamiferos fosiles. < Ib. I, 87—232, pl. 5—8. 1866. [Contains monographs of Gravigrada (pp. 149—182, pl. v) and Effodienta, a, Biloricata.—i. e. Glyptodontes (pp. 183—231, pl. vi—viii).]

—— Descripcion de cuatro especies de Delfines de la costa argentina. > Ib. I, 367—445, pl. xxi—xxviii, 1869. [Contains an anatomical monograph on Pontoporia Blainvillii, demonstrating its affinity with the Delphinidæ.]

—— Monografia de los Glyptodontes en el museo publico de Buenos Aires. < Ib. II, 1—107, pl. 1—12, 1870. [To be continued.]

CARUS (Julius Victor). Handbuch der zoologie von Jul. Victor Carus, . . . und C. E. A. Gerstaecker, ... Erster band. I. hälfte. Wirbelthiere, bearbeitet von J. Victor Carus.—Leipzig: Verlag von Wilhelm Engelmann, 1868.

[8vo., Bogen 1—27.—2⅓ th.] I. classe. Mammalia, [Säugethiere. pp. 39—191.]

COPE (Edward Drinkard). An addition to the extinct vertebrate fauna of the miocene period, with a synopsis of the extinct Cetacea of the United States. .... < Proceedings of the Academy of Natural Sciences of Philadelphia, 1867, 138—156.
[Cynorcidæ distinguished.]

DOYERE (M ··· P ···L ··· N ···). Notice sur un mammifère de Madagascar, formant le type d'un nouveau genre [Euplère] de la famille des carnassiers insectivores de M. Cuvier; par M. Doyère. < Annales des sciences naturelles ...., Seconde série. Tome quatrième. Zoologie. 1835, 270—283, pl. 8.

EDWARDS (Alphonse Milne). Recherches anatomiques, zoologiques et paléontologiques sur la famille des Chevrotains [Moschidæ et Tragulidæ]. .... < Annales des Sciences Naturelles. Cinquième série. Zoologie et Paléontologie. .... II, 1864, pp. 49—167, pl. 2—12.

—— Mémoire sur une nouvelle famille de l'ordre des Rongeurs [Lophiomides]. .... < Nouvelles Archives du Muséum d'Histoire Naturelle de Paris, III, 81—118, pl. 6—10, 1867.

EDWARDS (Henri Milne et Alphonse Milne). Recherches pour servir à l'histoire naturelle des Mammifères. .... Paris: Victor Masson et fils, ···, 1868 [—] 1870. [4to., liv. 1er—5er.—Chaque livr. 13 fr.]

FALCONER (Hugh). On the disputed affinities of the mammalian genus Plagiaulax, from the Purbeck beds. .... < The Quarterly Journal of the Geological Society of London, XVIII, 1862, 348—369.

—— Palæontological memoirs and notes of the late Hugh Falconer, A.M., M.D. .... With a biographical sketch of the author. Compiled and edited by Charles Murchison, M.D., F.R.S. .... [See "Contents."] London : Robert Hardwicke, ···· 1868. [8vo., 2 vols. (I,) lvi, 590 pp. 34 pl.; (II,) xiii, 675 pp. 38 pl.—42 sh.]

CONTENTS.
Vol. I. Fauna antiqua sivalensis.
" II. Mastodon, Elephant, Rhinoceros, Ossiferous caves, Primeval man and his cotemporaries.

FISCHER (Paul). Mémoire sur les cétacés du genre Ziphius, Cuvier. ···· < Nouvelles Archives du Muséum d'Histoire Naturelle de Paris, III, 41—79, pl. 4, 1867.
[Contains a synopsis of the Ziphioides.]

FLOWER (William Henry). Notes on the skeletons of whales in the principal museums of Holland and Belgium, with descriptions of two species apparently new to science. .... < Proceedings of the scientific meetings of the Zoological Society for the year 1864, 384—420.
—— Description of the skeleton of Inia geoffrensis and of the skull of Pontoporia blainvillii, with remarks on the systematic position of these animals

in the order Cetacea. . . . . < Transactions of the Zoological Society of London, VI, 87—116, pl. 4, 1867.
[Contains a systematic synopsis of the families and subfamilies Cetaceans.]

FLOWER (William Henry). On the affinities and probable habits of the extinct Australian marsupial, Thylacoleo carnifex, Owen. . . . . < The Quarterly Journal of the Geological Society of London, XXIV, 1868, 307—319.

—— On the value of the characters of the base of the cranium in the classification of the order Carnivora, and on the systematic position of Bassaris and other disputed forms. . . . . < Proceedings of the scientific meetings of the Zoological Society of London, for the year 1869, 4—37.

—— On the anatomy of the Proteles, Proteles cristatus (Sparrman). . . . . < 1b. 1869, 474—496, pl. 36.

GAUDRY (Albert). Animaux fossiles et géologie de l'Attique, d'après les recherches faites en 1855-56 et en 1860 sous les auspices de l'Académie des Sciences par Albert Gaudry. Paris : F. Sory, éditeur, 1862—1867. [4to., 474 pp. 1 l.; atlas 4 p. l., 1 map, 75 pl.—Published in 19 livr., at 6 fr. each.]

GEOFFROY SAINT-HILAIRE (Isidore). Muséum d'histoire naturelle de Paris.—Catalogue méthodique de la collection des Mammifères, de la collection des Oiseaux et des collections annexes. Par le professeur-administrateur M. Isidore Geoffroy Saint-Hilaire, · · · et les aides-naturalistes MM. Florent Prévost et Pucheran.—Paris : Gide et Baudry, ... 1851. [8vo. 3 p. l.— (Introduction.) xv. pp. — (Première partie.—Mammifères.—Catalogue des Primates, par M. Isidore Geoffroy Saint-Hilaire.) 1 p. l. vii, 96 pp.]

GERVAIS (Paul). Histoire naturelle des Mammifères avec l'indication de leurs mœurs, et de leurs rapports avec les arts, le commerce et l'agriculture . . . . [See "Contents."] Paris L. Curmer . . . . 1854 [—] 1855. [8vo., 2 v. (I) xxiv, 418 pp. 1 l. 18 col. pl., 14 uncol. pl.—21 fr.; (II) 2 p. l. 344 pp., 40 col. pl., 29 uncol. pl.—25 fr.]

CONTENTS.

1re partie. [Introduction, Primates, Cheiroptères, Insectivores, Rongeurs.] · · · 1854.

[2e partie.] Carnivores, Proboscidiens, Jumentés, Bisulques, Édentés, Marsupiaux, Monotrèmes, Phoques, Sirénides et Cétacés. . . . . 1855.

—— Animaux nouveaux ou rares recueillis pendant l'expedition dans les parties centrales de l'Amérique du Sud de Rio de Janeiro à Lima, et de Lima au Para ; exécutée par ordre du gouvernement français pendant les années 1843 à 1847, sous la direction du comte Francis de Castelnau. · · · Mammifères par M. Paul Gervais, · · · Paris, chez P. Bertrand, · · · 1855 [2 p. l., 116 pp. 20 pl.] < CASTELNAU (François de Laporte, comte de). Expedition dans les parties centrales de l'Amérique du Sud, de Rio de Janeiro à Lima, et de Lima au Para. 7e partie. Zoologie.

GIEBEL (Christian Gottfried Andreas). Die Säugethiere in zoologischer, anatomischer und palæontologischer beziehung umfassend dargestellt, ... Leipzig : verlag von Ambrosius Abel. 1855. [8vo., xii, 1108 pp.—7 r. th. 10 n. g.]

GILL (Theodore Nicholas). Prodrome of a monograph of the Pinnipedes.
. . . . < Communications of the Essex Institute. V, pp. 1—13, 1866.

——— The eared seals. [A review of memoir on the eared seals (Otariadæ), etc.,
by J. A. ALLEN.] < The American naturalist, a popular illustrated magazine
of natural history. IV, 675.

——— On the Sperm-whales [Physeteridæ], giant and pygmy. . . . . < Ib. IV,
725—743, 1871.

——— Synopsis of the primary subdivisions of the Cetaceans. . . . . < Communications of the Essex Institute. VI, 121—126, 1871.

GRAY (John Edward). Catalogue of the specimens of Mammalia in the collection of the British museum. Part III. Ungulata furcipeda. . . . . London :
printed by order of the trustees. 1852. [12mo., xvi, 286 pp. 37 pl.—12 sh.]

——— Catalogue of Seals [Pinnipedia] and Whales [Cete] in the British museum. . . . . Second edition. London : printed by order of the trustees. 1866.
[8vo., vii, 402 pp.—8 sh.]

——— Synopsis of the genera of Vespertilionidæ and Noctilionidæ. . . . . < The
Annals and Magazine of Natural History, · · · XVII. Third series, 1866, 89—93.

——— A revision of the genera of Rhinolophidæ, or horseshoe bats. . . . . < Proceedings of the scientific meetings of the Zoological Society of London for the
year 1866, 81—83.

——— Revision of the genera of Phyllostomidæ, or leaf-nosed bats. . . . . < Ib.,
1866, 111—118.

——— Catalogue of Carnivorous, Pachydermatous, and Edentate Mammalia in
the British museum. . . . . London : printed by order of the trustees. 1869.
[8vo., 4 p. l. 398 pp.—6 sh. 6 d.]

——— Catalogue of Monkeys, Lemurs, and Fruit-eating Bats in the collection of
the British museum. . . . . London : printed by order of the trustees. 1870.
[8vo., viii, 137 pp.]

HUXLEY (Thomas Henry). On the osteology of the genus Glyptodon. . . . .
< Philosophical Transactions of the Royal Society of London, CLV, 1865.
31—70, pl. 4—9.

——— On the structure of the stomach in Desmodus rufus. . . . . < Proceedings
of the scientific meetings of the Zoological Society of London for the year 1865.
386—390.

——— Reports of Professor Huxley's lectures on "The structure and classification of the Mammalia," delivered at the Royal College of Surgeons. < The
Medical Times and Gazette, 1864, I and II, viz :—
    Lecture I [—] IX. ANTHROPINI. [I, 153 : Distinctive characters and skele-
    ton. II, 177 : Muscles. III, 20 : Extremities. IV, 229 ; V, 256 : Brain.
    VI, 284 : Teeth and organs of reproduction. VII, 312 : Development.
    VIII, 343 ; IX, 369 : Variations and number of species.]
    Lecture X [—] XIX. ANTHROPOMORPHA. [(*Troglodytes niger.*) X, 398 ;
    XI, 428 ; XII, 456 : Skeleton and muscles. XIII, 486 : Larynx, Teeth,

Brain. XIV, 509 : Organs of reproduction, development, variations.——
Troglodytes gorilla XIV, 509; XV, 537; XVI, 564.——(*Simia satyrus.*)
XVI, 564; XVII, 595; XVIII, 617.——(*Hylobates*). XVIII, 617;
XIX, 647.——(*Characteristics of* ANTHROPOMORPHA). XIX, 647.]
Lecture XX [—] XXI. CYNOPITHECINA. (v. I,) p. 671. (v. II,) pp.
12 ; 40; 93; 123.
Lecture XXII [—] XXIII. PLATYRHINI. pp. 93; 123.
Lecture XXIII. ARCTOPITHECINI. p. 124.
Lecture XXIV. LEMURINI. CHEIROMYINI. Recapitulation. p. 145.

HUXLEY (Thomas Henry). Professor Huxley's Lectures at the Royal College
of Surgeons. [On Mammalia]. < The Lancet, 1866, I, viz :—
Lecture I [—] IV. SIRENIA. pp. 157—158; 180; 214—215; 289.
Lecture IV [—] IX. CETACEA. pp. 239; 268; 291; 324—325 · 350; 381.
Lecture X [—] XI. PINNIPEDIA. pp. 434—435; 465—466.
Lecture XII. DOG. p. 607.

LEIDY (Joseph). The extinct mammalian fauna of Dakota and Nebraska,
including an account of some allied forms from other localities, together with
a synopsis of the mammalian remains of North America. · · · ·. Preceded
with an introduction on the geology of the tertiary formations of Dakota and
Nebraska, by F. V. Hayden, M. D. Philadelphia, 1869. ═ Journal of the
Academy of Natural Sciences of Philadelphia, vol. VII, second series. Philadelphia : published for the Academy, by J. B. Lippincott & Co. 1869. [4to.,
472 pp., 30 pl., 1 map.—$20.]

LILLJEBORG (Wilhelm). Systematisk öfversigt af de Gnagande Däggdjuren,
Glires. · · · ·. Uppsala : Kongl. akad. bocktryckeriet, 1866. [4to., 1 p. l. 59 pp.
3 folded tables.]

McCOY (Frederick). On the species of Wombats [Phascolomyidæ]. (Abstract.)
· · · · < Transactions and Proceedings of the Royal Society of Victoria,
VIII, 266—270. 1868.

MARTIN (William Charles Linnæus). A general introduction to the Natural
History of Mammiferous Animals, with a particular view of the Physical
History of Man, and the more closely allied genera of the order Quadrumana,
or Monkeys, · · · ·. Illustrated with 296 anatomical, osteological, and other
incidental engravings on wood, and 12 full plate representations of animals,
drawn by William Harvey. London : Wright & Co. 1841. [8vo., 1 p. l.
545 pp., 12 pl.—16 sh.]

MIVART (St. George). Notes on the crania and dentition of the Lemuridæ.
· · · · < Proceedings of the Zoological Society of London, 1864, 611—648.

—— Contributions towards a more complete knowledge of the axial skeleton in
the Primates. · · · · < Ib., 1865, 545—592.
[Contains a synoptical arrangement of the order.]

—— On the structure and affinities of Microrhynchus laniger [Lemuridæ]. · · · ·
< Ib., 1866, 151.

—— On the skull of Indris diadema [Lemuridæ]. · · · · < Ib., 1867. 247.

—— Additional notes on the osteology of the Lemuridæ. · · · · < Ib., 1867, 960.

MIVART (St. George). Notes on the osteology of the Insectivora. . . . .— < The Journal of Anatomy and Physiology, I, 1867, 281—312 ; II, 1868, 117—154. [Contains a descriptive synopsis of the order.]

—— Notes sur l'ostéologie des Insectivores. . . . . < Annales des Sciences Naturelles. Cinquième série. Zoologie et paléontologie, VIII, 1867, 221—284 ; IX, 1868, 311—372. [A translation of the preceding.]

MORGAN (Lewis Henry). The American Beaver [Castoridæ] and his works. . . . . Philadelphia : J. B. Lippincott & Co. 1868. [8vo., 330 pp., 1 map, 23 pl.—$5.]

MURIE (James). On the saiga antelope, Saiga tartarica (Pall.) . . . . < Proceedings of the scientific meetings of the Zoological Society of London for the year 1870, 451—503.

—— Notes on the anatomy of the prongbuck, Antilocapra americana. . . . . < Ib. 1870, 334—368.

OWEN, F.R.S. (Richard). On the osteology of the Marsupialia. . . . . < Transactions of the Zoological Society of London, II, 1841, 379—408, pl. 68—71.

—— Outlines of a classification of the Marsupialia. . . . . < Ib., II, 1841. 315—333.

—— Description of the skeleton of an extinct gigantic sloth (*Mylodon robustus*, Owen), with observations on the osteology, natural affinities, and probable habits of the Megatherioid quadrupeds in general. By Richard Owen, F. R. S., Hunterian professor and conservator of the museum of the Royal college of surgeons in London. Published by direction of the council. London : · · · Sold by John Van Voorst, · · · · 1842. [4to., 176 pp., 24 pl. w. 24 expl. l.]

—— Description of some species of the extinct genus Nesodon, with remarks on the primary group (Toxodontia) of hoofed quadrupeds, to which that genus is referable. . . . . < Philosophical Transactions of the Royal Society of London. For the year MDCCCLIII. vol. 143, 291—310, pl. 15—18.

—— Palæontology or a systematic summary of extinct animals and their geological relations. . . . . Edinburgh : Adam and Charles Black. 1860. [8vo. xv, 420 pp.]

PETERS (Wilhelm Carl Hartwig). [22. Mai 1865.] IIr. W. Peters legte Abhandlungen zu einer monographie der Chiropteren vor und gab eine Übersicht der von ihm befolgten systematischen ordnung der hieher gehörigen gattungen. < Monatsberichte der königlichen Preuss. Akademie der Wissenschaften zu Berlin, 1865, 256—258.

—— [13. Juli 1865.] IIr. W. Peters las über flederthiere (*Vespertilio soricinus* Pallas, *Choroenycteris* Lichtenst., *Rhinophylla pumilio* nov. gen., *Artibeus fallax* nov. sp., *A. concolor* nov. sp., *Dermanura quadrivittatum* nov. sp., *Nycteris grandis* n. sp.). < Ib., 1865, 351—359. [Contains a synopsis of Stenoderminæ, pp. 356—359 ; continued on p. 524.]

PETERS (Wilhelm Carl Hartwig). [16. October 1865.] Hr. W. Peters las über die zu den Vampiri gehörigen fledderthiere und über die natürliche stellung der gattung Antrozous. < Ib., 1865, 503—524.

—— [22. Juni.] Hr. W. Peters las über die zu den Glossophagæ gehörigen fledderthiere und über eine neue art der gattung Colëura. < Ib., 1868, 361—386, 1 pl.

PICTET (Francois Jules). Traité de Paléontologie ou histoire naturelle des animaux fossiles considérés dans leurs rapports zoologiques et géologiques. .... Seconde édition, revue, corrigée, considérablement augmentée, accompagnée d'un atlas de 110 planches grand in-4°. ..... Paris, chez J.-B. Baillière, ... 1853 [—] 1857. [8vo. 4 v.; 4to. atlas.—80 fr.]

RÜTIMEYER (Ludwig). Versuch einer natürlichen geschichte des rindes, in seinen beziehungen zu den Wiederkauern im allgemeinen. [Eine anatomisch-palaeontologische monographie von Linné's genus *bos*. .... [4to., Erste abtheilung. 102 pp. 1 l. 2 pl.; Zweite abtheilung, 175 pp., 4 pl.] < Neue Denkschriften der allgemeinen schweizerischen Gesellschaft für die. gesammten Naturwissenschaften.—Nouveaux mémoires de la Société helvétique des sciences naturelles. XXIII, [Dritte dekade, II]. 1867.

SCLATER (Philip Lutley). Remarks on the arrangement and distribution of the Cervidæ. .... < Proceedings of the Zoological Society of London, 1870, 114—115.

TURNER (H ··· N ···, jun.). On the evidences of affinity afforded by the skull in the Ungulate mammalia. .... < Proceedings of the Zoological Society of London. Part XVII, 1849, 147—158.

VAN BENEDEN (Pierre Joseph). Recherches sur les Squalodons. .... < Mémoires de l'Académie royale de Belgique, XXXV, 1865.

WATERHOUSE (George R ···). A Natural History of the Mammalia. .... [See "Contents."] London: Hippolyte Baillière, ... 1846 [—] 1848 [8vo., (I), 3 p. 1. 553 pp., 22 pl. (11 col.).—; (II,) 1 pl. 500 pp., 22 pl. (11 col.).— each 29 sh; col., 34 sh 6 d.]

CONTENTS.

Vol. I. Containing the order Marsupiata, or Pouched animals, with 22 illustrations engraved on steel, and 18 engravings on wood.

Vol. II. Containing the order Rodentia, or Gnawing mammalia; with 22 illustrations engraved on steel, and [8] engravings on wood.

OF

CHARACTERS

OF THE SUB-DIVISIONS OF

MAMMALS,

WITH

A CATALOGUE OF THE GENERA.

———

WASHINGTON, D.C.
1871.

MAMMALS.

Abranchiate Vertebrates with a brain whose cerebral hemispheres are more or less connected (and in nearly inverse ratio) by an anterior commissure, and a superior transverse commissure (corpus callosum) ; the latter more or less roofing in the lateral ventricles: lungs and heart in the thorax, separated from the abdominal viscera by a muscular diaphragm : aorta single and reflected over the left bronchus : blood with red non-nucleated blood-corpuscles ; undergoing a complete circulation, being entirely received and transmitted by the right half of the quadrilocular heart to the lungs for aeration, (and warmed,) and afterwards returned by the other half through the system. Skull with two condyles, chiefly developed on the exoccipital elements, (one on each side of the foramen magnum): with the malleus and incus superadded as specialized auditory ossicles : and the lower jaw (composed of a pair of simple rami) articulated directly by convex condyles with the squamosal bones. Dermal appendages developed as hairs. Viviparous: foetus developed from a minute egg: young nourished after birth by a fluid (milk) secreted in peculiar glands (mammary) by the mother.

SUB-CLASSES.

I. Brain with superior transverse commissure composed of a body as well as psalterial fibres ; and with a well developed septum. Sternum with no element in front of the manubrium or presternum. Coracoid not connected with the sternum, but early anchylosed with and developed as a simple process of the scapula. Oviducts debouching into a double or single vagina, (and not into a common cloacal chamber). Testes variable in position, but the vasa deferentia open directly or indirectly into a distinct and complete urethra, (and not into a cloacal cavity). Ureters discharge directly into the bladder the renal secretion, which thence passes into the urethra. Mammary glands with well developed nipples.

A. Brain with the cerebral hemispheres connected by a more or less well-developed corpus callosum and a reduced anterior commissure. Vagina a single tube, but sometimes with a partial septum. Young retained within the womb till of considerable size and nearly perfect development, and deriving its nourishment from the mother through the intervention of a "placenta" (developed from the allantois) till birth. Scrotum never in front of penis.

MONODELPHIA. (I.)

B. Brain with the cerebral hemispheres chiefly connected by a well-developed anterior commissure, the corpus callosum being rudimentary.

Vagina more or less completely dividing into two separate passages. Young born when of very small size and imperfect development; never connected by a placenta with the mother, but when born attached by her to the nipple, from which the milk is forced by herself into the mouth of the young. Scrotum in front of penis.

DIDELPHIA. (II.)

II. Brain with the superior transverse commissure with no well defined psalterial fibres; and with the septum very much reduced in size. (Flower). Sternum with a peculiar T-shaped bone (the episternum or interclavicle) in advance of the manubrium or presternum. Coracoid extending from the clavicle to the sternum, and only towards maturity anchylosed with the scapula. The oviducts, enlarged below into uterine pouches, but opening separately from one another, as in oviparous vertebrates, debouch, not into a distinct vagina, but into a cloacal chamber, common to the urinary and genital products, and to the fæces. Testes abdominal in position throughout life, and the vasa deferentia open into the cloaca, and not into a distinct urethral passage. Ureters pour the renal secretion, not into the bladder, which is connected with the upper extremity of the cloaca, but into the latter cavity itself. Mammary glands with no distinct nipples. (Huxley.)

ORNITHODELPHIA. (III.)

I. MONODELPHIA

ORDERS.

I. Brain with a relatively large cerebrum, behind overlapping much or all of the cerebellum, and in front much or all of the olfactory lobes : corpus callosum (attypically) continued horizontally backwards to or beyond the vertical of the hippocampal sulcus, developing in front a well-defined recurved rostrum.

SUPER-ORDER EDUCABILIA.

A. Posterior members and pelvis well developed. Periotic and tympanic bones articulated with the squamosal; (etypically, free and otherwise modified, e. g. Tapiridae).

1. Legs almost or entirely exserted outside of the common abdominal integument. First digit (great toe) of hind foot (pes) enlarged, opposable to the others, (exceptionally resuming parallelism with them,) furnished with a nail. Brain with a well-developed calcarine sulcus, giving rise to a hippocampus minor within the posterior cornu of the ventricle by which the posterior lobe of the cerebrum is traversed. (Flower.) (Incisors four in each jaw; etypically, two—or all in upper jaw—suppressed. Clavicles completely developed.)

a. Digits with corneous appendages developed as claws (i. e. compressed) or, attypically, as nails (i. e. depressed). Teeth of three kinds, (canines of second set etypically atrophied,) all encased in enamel ; (molars mostly two- or three-rooted). Placenta deciduate, discoidal.

PRIMATES. (I.)

2. Legs with the proximal joints (humerus and femur) more or less inclosed in the common abdominal integument. First digit of hind foot attypically reduced or atrophied ; etypically hypertrophied (e. g. *Pinnipedia*). Brain with no calcarine sulcus. (Incisors archetypically six in each jaw; etypically, two or more suppressed. Clavicles rudimentary, or—in (b) Ungulate Series—none.)

a. Digits with corneous appendages developed as claws. Teeth of three kinds, all encased with enamel: canines specialized and robust: (molars mostly two- or three-rooted—etypically one-rooted,—attypically one $\left(\frac{\text{p.m. } 4.}{\text{m. } 1.}\right)$ or more in each jaw sectorial, followed by tubercular ones.) Scaphoid and lunar consolidated into one bone. Placenta deciduate, zonary.

FERAE. (II.)

b. Digits with corneous appendages developed as hoofs. Teeth of three kinds, (canines and incisors of second set exceptionally in part undeveloped,) all encased in enamel: (molars attypically two- or three-rooted, attypically with grinding surfaces.) Scaphoid and lunar separate. Placenta diversiform.

b. 1. Incisors (archetypically $\frac{6}{6}$; often, especially in the upper jaw, reduced in number or wholly suppressed; implanted by simple roots,) with incisorial crowns. Feet with inferior (or, rather, posterior) surfaces with a hairy skin continuous with the rest of the integument: carpal bones in two interlocking rows; cuneiform narrow and affording a diminished surface of attachment forwards for the ulna (which is retrorse beside the radius); unciform and lunar articulating with each other and interposed between the cuneiform and magnum: hind foot with the astragalus at its anterior portion scarcely deflected inwards, articulating more or less with the cuboid as well as navicular: toes (not more than four completely developed) with terminal joints encased in thick hoofs. Placenta non-deciduate, (diffuse or cotyledonary).

UNGULATA. (III.)

b. 2. Incisors ($\frac{6}{6}$ or $\frac{4}{4}$; variable as to insertion,) with incisorial crowns. Feet mostly unknown: carpal bones unknown: hind foot with the astragalus at its anterior portion inclined obliquely inwards, articulating in front only with the navicular. Calcaneum with an extensive upwards surface for the articulation of the fibula, and with a large lateral process articulating in front with the astragalus. Molars of upper jaw broad and extending into an externo-anterior angle; of lower jaw, narrow and continuous in a uniform row.

TOXODONTIA. (IV.)

b. 3. Incisors (¹) of upper jaw next to symphysis (with persistent pulps) long and curved; those of lower jaw straight and normal. Feet with inferior surfaces furnished with pads, (as in Rodents and Carnivores): carpal bones in two interlocking rows; cuneiform extending inwards, (and articulating with magnum,) and affording an enlarged surface of attachment forwards for the ulna (which is antrorsely produced); unciform and lunar separated by the interposition of the cuneiform and magnum: hind foot with the astragalus at its anterior portion extended and, as a whole, much deflected inwards, articulating in front only with the navicular: toes (four to the front feet, three to the hind) with terminal.phalanges encased in hoofs; (inner nail of hind foot curved). Placenta deciduate, zonary.

HYRACOIDEA. (V.)

b. 4. Incisors (³⁄₀, or, in extinct forms, ³⁄₂ or ⁰⁄₂, renewed from persistent pulps,) developed as long tusks curved outwards. Feet with palmar and plantar surfaces invested in extended pad-like integuments, which also underlie the toes: carpal bones in two regular (not interlocking) rows, broad and short; cuneiform extended inwards—broad, and furnishing an enlarged surface of attachment forwards for the ulna (which is antrorsely produced). Unciform directly in front of cuneiform, and magnum directly in front of lunar: hind foot with the astragalus at its anterior portion very short, (convex,) and not deflected inwards, articulating in front only with the navicular: toes (five to each foot, in known forms,) encased in broad shallow hoofs. Placenta deciduate, zonary.—Snout produced into a very long proboscis. Legs mostly exserted outside the abdominal integument; and with the proximal and succeeding joints extensible in the same line.

PROBOSCIDEA. (VI.)

B. Posterior members and pelvis more or less completely atrophied; the form of the body being fish-like, furnished with a horizontal tail, and specialized for progression in the water. Periotic and tympanic bones anchylosed together, but not articulated with the squamosal.

1. Brain narrow. Skull with the foramen magnum posterior, directed somewhat downwards: supra-occipital nearly vertical and not extending forwards, the parietals meeting and interposed between it and the frontals. Periotic with a posterior irregularly rounded part; tympanic annuliform. Lower jaw with well-developed ascending rami and normal transverse condyles and coronoid processes. Lateral teeth molar, and adapted to trituration of herbage. Neck moderate; second

cervical vertebra with an odontoid process. Anterior members moderately long, flexed at the elbow; with carpal bones and phalanges directly articulated with the adjoining ones; and with normal digits. Mammæ two, pectoral.—Heart deeply fissured between the ventricles.

SIRENIA. (VII.)

2. Brain broad. Skull with the foramen magnum entirely posterior, directed somewhat upwards: supra-occipital very large, sloping forwards, and (attypical'y) extending forwards over or between the frontals. Periotic attenuated backwards; tympanic solid, entire. Lower jaw with no ascending ramus, with its narrow condyles at the posterior extremities or angles of the rami, and with only rudimentary coronoid processes. Teeth conic or compressed, monophyodont. Neck attypically very short; second cervical vertebræ with no odontoid process. Anterior members (attypically) abbreviated, extended backwards in a continuous line; with carpal bones and phalanges often separated by cartilage; and with the second digit composed of more than three phalanges. Mammæ two, inguinal.

CETE. (VIII.)

II. Brain with a relatively small cerebrum, leaving behind much of the cerebellum exposed, and in front much of the olfactory lobes: corpus callosum extending more or less obliquely upwards and terminating before the vertical of the hippocampal sulcus; with no well defined rostrum in front.

SUPER-ORDER INEDUCABILIA.

A. Teeth encased in enamel: incisors (very variable as to number) without persistent pulps: canines present (but sometimes modified in form): molars attypically with sharp and pointed cusps. Lower jaw with condyles transverse, received into special glenoid sockets. Placenta discoidal deciduate.

1. Anterior members adapted for flight: the ulna and radius being united, and the metacarpal bones and phalanges—2 to 5—much elongated; the whole sustaining a very thin leathery skin arising from the sides of the body, and extending backwards on the hind members, down to their tarsi. Mammæ pectoral.

CHIROPTERA. (IX.)

2. Anterior as well as posterior members adapted for walking or grasping: the ulna and radius entirely or partly separated: metacarpal bones and phalanges normally developed. Mammæ abdominal: (etypically—in Dermoptera, &c.—pectoral).

July, 1871.

INSECTIVORA. (X.)

B. Teeth encased in enamel: incisors ($\frac{2}{2}$; exceptionally, also two supplementary posterior teeth,) continually reproduced from persistent pulps, and growing in a circular direction: canines none: molars attypically with ridged surfaces. Lower jaw with condyles longitudinal, and not received in special glenoid sockets, but gliding freely backwards and forwards in longitudinal furrows. Members and feet ambulatorial. Placenta discoidal deciduate.

GLIRES. (XI.)

C. Teeth (when developed) not encased in enamel: incisors typically absent (lateral present in *Dasypus*): molars variable: members and feet ambulatorial, (modified often for grasping and digging). Placenta diversiform (discoidal deciduate in *Orycteropodidæ* and *Dasypodidæ*; diffuse deciduate in *Manididæ*; and coyledonous non-deciduate? in *Bradypodidæ*).

BRUTA. (XII.)

I. PRIMATES.

SUB-ORDERS.

I. Cerebrum with its posterior lobe much developed, wholly or mostly covering the cerebellum. Skull with lachrymal foramen within the orbit. Orbit separated from temporal fossa by the union of the alisphenoid and malar bones. Ears rounded, each with a distinct lobule. Female with uterus undivided, and clitoris imperforate. Mammæ (2) exclusively pectoral.

ANTHROPOIDEA.

II. Cerebrum with the posterior lobe not extended backwards over the entire cerebellum, a considerable portion of the latter being uncovered. Skull with lachrymal foramen outside the orbit. Orbits open behind, (partially closed in Tarsiidae). Ears more or less produced upwards and pointed, angulated at their extremities, with no distinct lobules. Female with uterus two-horned, and the clitoris perforated by the urethra. Mammæ variable.

PROSIMIAE.

ANTHROPOIDEA.

FAMILIES.

I. Fore limbs withdrawn completely from the *locomotive* series, and transferred to the *cephalic* (Dana). Form habitually erect, except in infancy. Feet with the great toe produced, and in same plane with others. Teeth in an uninterrupted series. Hair scant. (*Bimana*.)

HOMINIDAE. (I.)

II. Fore limbs more or less employed in progression. Form prone, exceptionally erect. Feet with the great toe more or less abbreviated, thumb-like, and opposable to the others. Dental series interrupted by diastemas, especially in the upper jaw between canines and incisors. Hair dense. (*Simiæ.*)

A. A bony external auditory meatus well developed, at the bottom of which is the membrana tympani. Pre-molars $\frac{2}{2} \times 2$. (Teeth M $\frac{3}{3}$ PM $\frac{2}{2}$ C $\frac{1}{1}$ I $\frac{2}{2} \times 2$.) Nose with the median septum thin and narrow (exceptionally, broad), and the nostrils correspondingly approximated. (*Simiæ catarrhinæ.*)

1. Spinal column with a slight sigmoid curve; lumbar as well as dorsal neural spines directed more or less backwards. Sacrum large and solid, composed of four vertebræ tapering gradually backwards. Sternum broad and short, with three or four bones between the manubrium and xiphoid cartilage. Anterior limbs much longer than posterior.

SIMIIDAE. (II.)

2. Spinal column with a simple curve; neural spines of lumbar and last dorsal vertebræ inclined forwards. Sacrum moderate, composed generally of three vertebræ not tapering gradually. Sternum elongated and narrow. Anterior limbs shorter than posterior; rarely elongated.

CYNOPITHECIDAE. (III.)

B. Bony external auditory meatus null, and the tympanic membrane attached to a ring close to the surface. Pre-molars $\frac{3}{3} \times 2$. Nose with the septum broad and flattened (exceptionally, narrow), and the nostrils proportionally distant. (*Simiæ platyrhinæ.*)

1. Teeth (M $\frac{3}{3}$ PM $\frac{3}{3}$ C $\frac{1}{1}$ I $\frac{2}{2} \times 2 =$) 36. Manus with inner digit (when developed) more or less slightly opposable to the rest.

CEBIDAE. (IV.)

2. Teeth (M $\frac{2}{2}$ PM $\frac{3}{3}$ C $\frac{1}{1}$ I $\frac{2}{2} \times 2 =$) 32. Manus with inner digit not opposable, but on same plane as rest; all armed with elongated compressed claws.

MIDIDAE. (V.)

I. HOMINIDAE.

Single genus.

Homo.

II. SIMIIDAE.

SUB-FAMILIES.

I. Form robust. Ilia broad, alate. Cerebrum projecting backwards over the cerebellum. Buttocks without callosities.

SIMIINAE. (A.)

II. Form slender. Ilia narrow, not alate. Cerebrum scarcely or not projecting backwards over the cerebellum. Buttocks with callosities.

HYLOBATINAE. (B.)

Gorilla I. Geoff.
Mimetes Leach=*Troglodytes*, Geoff.=*Anthropopithecus*, Bl.
Simia Linn.=*Pithecus* Geoff.

B. HYLOBATINAE.

Siamanga Gray.
Hylobates Ill.

Extinct Simiidæ?

Pliopithecus Gerv.
Dryopithecus Lartet.

III. CYNOPITHECIDAE.

SUB-FAMILIES.

I. Stomach complex; the cardiac portion dilated; the pyloric elongated.
Cheek pouches obsolete.

SEMNOPITHECINAE. (A.)

II. Stomach simple, as in man. Cheek pouches developed.

CYNOPITHECINAE. (B.)

A. SEMNOPITHECINAE.

Nasalis Geoff.
Lasiopyga Ill., Gray.
Semnopithecus F. Cuv.
Colobus Ill.
Guereza Gray.

B. CYNOPITHECINAE.

§. 1.

Miopithecus I. Geoff.
Cercopithecus Erxl.
   Cercopithecus sensu strict.   *Chlorocebus* Gray.

§. 2.

Cercocebus Geoff.
Macacus Lac., Desm.
   Macacus sensu strict.   *Silenus* Gray.
Inuus Geoff.
Theropithecus I. Geoff.=*Gelada* Gray.
Cynopithecus I. Geoff.

§. 3.

Papio Erxl., Cuv., Geoff.=*Cynocephalus* Lac.
   Cynocephalus, sensu strict.   *Hamadryas* Less., Gray.
Mandrilla Cuv.,=*Mormon* Less.
   Mormon, Gray, not Ill.   *Choeropithecus* Gray.

Extinct Cynopithecidæ.

Mesopithecus Gaudry.
Coenopithecus Rutimeyer.

IV. CEBÌDAE.

SUB-FAMILIES.

I. Cerebrum with posterior lobe abbreviated, scarcely covering the cerebellum behind. Hyoid bone and thyroid cartilage greatly developed : hyoid bone expanded into a, sub-globular drum, with thin osseous walls, the larger cornua projecting backwards, the lesser obsolete. Incisors vertical.

MYCETINAE. (A.)

II. Cerebrum with posterior lobe enlarged, extending backwards much beyond the cerebellum. Hyoid bone and thyroid cartilage moderate.

A. Incisors vertical.

1. Cerebrum with convolutions well marked. Tail more or less prehensile.

CEBINAE. (B.)

2. Cerebrum with convolutions obsolete. Tail not prehensile.

NYCTIPITHECINAE. (C.)

B. Incisors inclined forwards. Tail more or less abbreviated and bushy.

PITHECIINAE. (D.)

A. MYCETINAE.

Aluatta Lac.=*Mycetes* Ill.

B. CEBINAE.

§. 1.

Cebus Erxl.

§. 2.

Sapajou Lac.= *Ateles* Geoff.
Eriodes I. Geoff.=*Brachyteles* Gray.
Lagothrix Geoff.

C. NYCTIPITHECINAE.

§. 1.

Nyctipithecus Spix=*Aotus* (Humb) Ill. (Inapplicable.)

§. 2.

Callithrix Geoff.
Saimiris Geoff., Gerv.= *Chrysothrix* Wagn.

D. PITHECIINAE.

Pithecia Desm.
Chiropotes Ill., Gray.
Brachyurus Spix=*Ouakaria* Gray.

Extinct Cebidæ.

Protopithecus Lund.

V. MIDIDAE.

Saguinus Lac=*Hapale* Geoff.
  Hapale Gray.
  Cebuella Gray.
Midas Geoff.
  Leontopithecus Gray.
  Midas Gray.

Jacchus Gray.
Mico Gray.

Œdipus Gray.
Seniocebus Gray.

PROSIMIAE.

FAMILIES.

I. Teeth of three kinds, the canines being retained through life. Incisors small, with simple roots. Pectoral mammæ developed, in addition to inguinal ones. (Owen.) (*Lemuroidea.*)

A. Fibula entirely distinct from the tibia. Skull with the orbits open behind. Incisors of upper jaw small, (rarely wanting,) separated into two groups by a symphysial interspace; of lower jaw, larger, contiguous, and proclivous; canines of lower jaw parallel with and like incisors. Pes with the second toe armed with a subulate claw; rest with flattened nails.

LEMURIDAE. (VI.)

B. Fibula partially anchylosed with the tibia. Skull with the orbits partially closed behind by the union above of the alisphenoid with the jugal. Incisors of upper jaw (4) contiguous, inner large and conic; of lower (2) contiguous and opposed to large upper teeth: canines of lower jaw normal. Pes with the second and third toes armed with subulate nails; rest with flattened pointed nails.

TARSIIDAE. (VII.)

II. Teeth of two kinds, the canines being early deciduous. Incisors $\frac{2}{1}$, gliriform, continually reinforced from the formative pulp; the fangs very long, those of the lower jaw extending backwards to the base of the coronoid processes. Inguinal teats only developed.—Manus with the middle finger very attenuated, and provided with a narrow scooped nail; rest of nails (except of thumb of pes) similar, subulate. (*Daubentonioidea.*)

DAUBENTONIIDAE. (VIII.)

SUPER-FAMILY LEMUROIDEA.

VI. LEMURIDAE.

SUB-FAMILIES.

I. Teeth 30; i. e. M $\frac{3}{3}$ P.M. $\frac{2}{2}$ C. $\frac{1}{1}$ I. $\frac{2-}{1}$×2.

INDRISINAE. (A.)

II. Teeth 36 (exceptionally 32); i. e. M. $\frac{3}{3}$ P.M. $\frac{3}{3}$ C. $\frac{1}{1}$ I. $\frac{2-}{2}$×2 (exceptionally, I. $\frac{0}{1}$ in adult).

A. Tarsus short or of moderate length.

1. Hind limbs considerably longer than the fore. Neural spines of last dorsal and lumbar vertebræ inclined forwards. Ears (in typical forms) moderate, with the anterior portion of the helix well developed, folded over the fossæ of the concha and antihelix, and with the tragus and antitragus distinct. Tail elongated, not less than two-thirds the length of body.

LEMURINAE. (B.)

2. Hind and fore limbs sub-equal, or fore ones shorter. Neural spines of dorsal and lumbar vertebræ inclined backwards. Ear (in typical forms) small, with the helix little marked, and tragus and antitragus obsolete. Tail short (always shorter than half the length of the body), rudimentary, or absent. (Mivart.)

NYCTICEBINAE. (C.)

B. Tarsus very long; calcaneum more than one-third the length of the tibia; naviculare much longer than the cuboid (Mivart).—Hind limbs much longer than the fore. Neural spines of the twelfth or thirteenth dorsal vertebræ turned forwards. Ear very large, with the pinna prolonged upwards.

GALAGININAE. (D.)

A. INDRISINAE.

Indris Cuv. Geoff.=*Lichanotus* Ill.=*Pithelemnr* Less.
Propithecus Benn.=*Macromerus* Smith.
Microrhynchus Jourd.=*Avahis* Geoff.

B. LEMURINAE.

Lemur Linn.
  Varecia Gray.   *Prosimia* Gray.
  Lemur Gray.
Hapalemur I. Geoff.
  Hapalemur Gray.   *Prolemur* Gray.
Lepilemur I. Geoff.
Chirogaleus Geoff. (St. G. Mivart.)=*Myspithecus* F. Cuv.

C. NYCTICEBINAE.
§. 1.

Perodicticus Benn.
Arctocebus Gray, Huxl.

§. 2.

Nycticebus Geoff.=*Stenops* Ill.=*Bradylemur* (Blainv.) Less.
Loris Geoff.=*Arachnocebus* Less.

D. GALAGININAE.

Galago Geoff.=*Otolicnus* Ill.
  Otolemur Coq.=*Callotus* Gray.   *Otogale* Gray.
  Galago sensu strict.   *Hemigalago* Dahlb.
Microcebus Geoff.
  Murilemur Gray.   *Phaner* Gray.
  Azema Gray.   *Mirza* Gray.

VII. TARSIIDAE.

Tarsius Storr= *Macrotarsus* C. & G.= *Cephalopachus* Sw.= *Hypsicebus* Less.

SUPER-FAMILY DAUBENTONIOIDEA.

. VIII. DAUBENTONIIDAE.

Daubentonia Geoff.= *Aye-aye* Lac.= *Cheiromys* Cuv.

FERAE.

SUB-ORDERS.

I. Body more or less raised, with the legs exserted beyond the elbows and knees, and with the feet (generally with free toes) adapted for walking. Manus and pes with first phalanges and digits not enlarged nor produced beyond the others (attypically more or less reduced or even atrophied). Skull moderately compressed between the orbits: with a distinct lachrymal bone, perforated by a canal (the lachrymal), and more or less exserted outside the orbit, and, in conjunction with the malar, forming the anterior margin of the orbit: palatines extending forwards laterally between the frontal and maxillary bones, and leaving no vacuity. Tympanic bounded behind by the exoccipital. Deciduous dentition well developed.

FISSIPEDIA.

II. Body prone, with the legs confined in the common integument beyond the elbows and knees, (with the feet rotated backwards, and with toes connected together), and especially adapted for swimming. Manus and pes with first phalanges and digits enlarged and produced beyond the others. Skull much compressed between the orbits: with a lachrymal bone early united with the maxillary, imperforate, and entirely contained within the orbit: malar applied to the inner side of a transverse zygomatic process of the maxillary and not continued to the front of the orbit (which is therefore bounded by the maxillary): palatines not extending forwards laterally, extensive vacuities intervening between the frontal and maxillary bones. Tympanic separated from the exoccipitals by a vacuity as well as by the re-entering periotic bones. Deciduous dentition much reduced and rudimentary.

PINNIPEDIA.

FISSIPEDIA.

FAMILIES.

I. Skull with the paroccipital process applied closely to the auditory bulla; the mastoid process small or obsolete; external auditory meatus very short or imperfect. Intestinal canal provided with a cœcum. Prostate gland salient.

A. Skull with carotid canal minute and superficial or obsolete; condyloid foramen and foramen lacerum posticum debouching in a common fossa: glenoid foramen minute or null. Os penis rudimentary (in *Cryptoproctidæ*, enlarged). Cowper's glands present. (*Aeluroidea.*)

1. Teeth 28–30 (M $\frac{1}{1}$, PM $\frac{3}{2}$ or $\frac{2}{2}$, C $\frac{1}{1}$, I $\frac{3}{3}\times2$): true molar of upper jaw small, tubercular; of lower, sectorial. Snout very short, decurved. Bulla divided by a septum into posterior and anterior chambers communicating with each other by a narrow aperture. (*Aeluroidea typica.*)

   a. Body compact. Feet digitigrade, with the palms and soles hairy. Skull with no alisphenoid canal.

FELIDAE. (IX.)

   b. Body elongated, viverriform. Feet plantigrade, with the palms and soles bald. Skull with a distinct alisphenoid canal.

CRYPTOPROCTIDAE. (X.)

2. Teeth 32–34, diversiform, but no tubercular (or second true) molar in lower jaw. Snout moderate, depressed. Bulla with no septum. Feet digitigrade. (*Aeluroidea hyæniformia.*)

   a. Teeth 32 (M $\frac{1}{1}$? PM $\frac{3-4}{3}$? C $\frac{1}{1}$, I $\frac{3}{3}\times2$); molars very small and distant; no functionalized sectorial molars.

PROTELIDAE. (XI.)

   b. Teeth 34 (M $\frac{1}{1}$, PM $\frac{4}{3}$, C $\frac{1}{1}$, I $\frac{3}{3}\times2$); molars large and approximated; true molar of upper jaw reduced, tubercular; last pre-molar sectorial, feline: true molar of lower jaw sectorial.

HYÆNIDAE. (XII.)

3. Teeth 36–40 (M $\frac{2}{2}$—rarely $\frac{1}{2}$—PM $\frac{4}{4}$—exceptionally $\frac{3}{3}$—C $\frac{1}{1}$, I $\frac{3}{3}\times2$); true molars of the upper, and last of the lower jaw tubercular. Snout moderate or elongated, depressed. Auditory bulla divided. (*Aeluroidea viverriformia.*)

   a. Skull irregularly flattened behind above foramen magnum; with the snout moderate or robust. Incisors approximated; canines robust.

VIVERRIDAE. (XIII.)

   b. Skull convex behind above foramen magnum (at least, especially so in young); with the snout slender. Incisors not approximated; canines small.

EUPLERIDAE. (XIV.)

B. Skull with the carotid canal well developed, but opening into the foramen lacerum posticum; condyloid foramen distinct; glenoid foramen patent. Os penis large. Cowper's glands not developed. (*Cynoidea.*)

1. Teeth typically 42; varying between 38 and 46 (the true molars being the varying element.—M $\frac{2}{3}$ ($\frac{1}{2}-\frac{3}{4}$), PM $\frac{4}{4}$, C $\frac{1}{1}$, I $\frac{3}{3}\times2$).

CANIDAE. (XV.)

II. Skull with the paroccipital process not closely applied to the auditory bulla; the mastoid process prominent and projecting outwards or downwards behind the external auditory meatus; external auditory meatus diversiform. Intestinal canal with no cœcum. Prostate gland not salient, being contained in the thickened walls of the urethra.—Skull with the carotid canal distinct, and more or less in advance of the foramen lacerum posticum; condyloid foramen also distinct from the foramen lacerum posticum; glenoid foramen generally well defined. Os penis very large. Cowper's glands not developed.—(*Arctoidea.*)

A. True molars of upper jaw one (M $\frac{1}{2}$; rarely—in *Mellivorinæ*—$\frac{1}{1}$); last pre-molar of upper jaw sectorial (rarely—in *Enhydrinæ*—with blunt tubercles). (*Arctoidea musteliformia.*)

MUSTELIDAE. (XVI.)

B. True molars of upper jaw two; last pre-molar of upper jaw tubercular (rarely—in *Bassarididæ*—sectorial).

1. Last molar of upper jaw oblong and exceeding the first: three true molars in lower jaw; first narrowest but longest; second oblong and broader. Foramen lacerum posticum introrse, behind the postero-internal angle of the tympanic bone; carotid canal little in advance of the foramen lacerum posticum. Tail rudimentary. (*Arctoidea typica.*)

URSIDAE. (XVII.)

2. Last molar of upper jaw more or less transverse and compressed forwards; two true molars in lower jaw; first broadest. Foramen lacerum posticum antrorse from postero-internal angle of the tympanic bone; carotid canal nearly at or in advance of middle of inner wall of the auditory bulla. Tail well developed. (*Arctoidea procyoniformia.*)

a. Alisphenoid canal developed: auditory bulla very small, and with a very prolonged bony floor to the auditory meatus: paroccipital process long and trigonal, standing backwards and outwards, quite unconnected with the bulla. (Flower.)—Teeth 36 (M $\frac{2}{2}$, PM $\frac{3}{3}$, C $\frac{1}{1}$, I $\frac{3}{3}$×2).

AELURIDAE. (XVIII.)

b. Alisphenoid canal none: auditory bulla well developed, and with a short bony floor to the auditory meatus: paroccipital process short and blunt, somewhat hooked, generally contiguous to the bulla at the base.

b. 1. Teeth 36 (M $\frac{2}{2}$, PM $\frac{3}{3}$, C $\frac{1}{1}$, I $\frac{3}{3}$×2); last pre-molar of upper jaw and first molar of lower tubercular. Snout abbreviated, decurved. Lower jaw very stout, with an extensive anchylosed symphysis, with high coronoid processes, and extended backwards and downwards at the angles.

CERCOLEPTIDAE. (XIX.)

b. 2. Teeth 40 (M $\frac{2}{2}$, PM $\frac{4}{4}$, C $\frac{1}{1}$, I $\frac{3}{3}$×2); last pre-molar of upper jaw and first molar of lower tubercular. Lower jaw moderate or slender, with a reduced symphysis, with recurved coronoid processes, and extended upwards to the angles, which are near the condyles.

PROCYONIDAE. (XX.)

b. 3. Teeth 40 (M $\frac{2}{2}$, PM $\frac{4}{4}$, C $\frac{1}{1}$, I $\frac{3}{3}\times2$), resembling those of *Canidæ*; first upper pre-molars sometimes deciduous; last pre-molar of upper jaw and first molar of lower sectorial. Lower jaw as in *Procyonidæ*.

BASSARIDIDAE. (XXI.)

Familiæ incertæ sedis.

1. Teeth 32? (M $\frac{2}{2}$? PM $\frac{3}{3}$? C $\frac{1}{1}$? I $\frac{3}{3}$;$\times2$)? last pre-molar of lower jaw moderate; first molar obtusely sectorial; second oblong, tuberculated.

SIMOCYONIDAE. (XXII.)

2. Teeth 44 (M $\frac{3}{3}$, PM $\frac{4}{4}$, C $\frac{1}{1}$, I $\frac{3}{3}\times2$)? last pre-molar of *upper* jaw tri-tuberculate; true molars tuberculate.

ARCTOCYONIDAE. (XXIII.)

3. Teeth 44? (M $\frac{3}{3}$, PM $\frac{4}{4}$, C $\frac{1}{1}$, I $\frac{3}{3}\times2$)? last pre-molar of *lower* jaw enlarged; first as well as second and third molars sectorial.

HYÆNODONTIDAE. (XXIV.)

SUPER-FAMILY AELUROIDEA.

IX. FELIDAE.

SUB-FAMILIES.

I. Canine teeth of upper jaw moderate, with transversely convex anterior and posterior margins; those of lower jaw equal to upper and much exceeding adjoining incisors.

A. Sectorial tooth of upper jaw with an inwardly projecting antero-internal lobe. Claws retractile.

FELINAE. (A.)

B. Sectorial tooth of upper jaw with no internal lobe. Claws not retractile.

GUEPARDINAE. (B.)

II. Canine teeth of upper jaw enormously developed, compressed, and with distal trenchant anterior and posterior margins; those of lower jaw reduced in inverse ratio, and not much larger than the adjoining incisors. (Sectorial tooth of upper jaw with a transverse inner lobe some distance in front of the anterior end of the tooth.)

MACHAERODONTINAE. (C.)

A. FELINAE.

Lynx Raf.	
Lyncus Gray.	*Caracal* Gray.
Neofelis Gray.	
Viverriceps Gray.	
Felis Linn.	
Uncia Gray.	*Leo* Gray.
Tigris Gray.	*Leopardus* Gray.
Pardalina Gray.	*Catolynx* Gray.

Pajeros Gray.
Chaus Gray.
Aelurina Gerv.=*Ailurogale* Fitz.

Felis Gray.

B. GUEPARDINAE.

Gueparda Gray.

C. MACHAERODONTINAE.

All extinct.

Drepanodon Nesti, Bronn.
Macherodus Kaup, Bronn.
Smilodon Lund, Bronn.

Other Felidae of Extinct Genera.

Pseudælurus Gerv.
Trucifelis Leidy.
? *Dinictis* Leidy.
? *Aelurodon* Leidy.

X. CRYPTOPROCTIDAE.

Single Genus.

Cryptoprocta Bennett.

XI. PROTELIDAE.

Single Genus.

Proteles I. Geoff.

XII. HYAENIDAE.

Genera.

Hyaena Linn.
Crocuta Gray.

XIII. VIVERRIDAE.

SUB-FAMILIES.

I. Auditory bulla divided by an oblique groove into two portions; an anterior with the auditory meatus, and a posterior more inflated and larger portion. (Flower.) Toes short, regularly arched; the last phalanges bent up, withdrawing the claws into a sheath; claws sharp. (Gray, s. *Aeluropodae.*)

   A. Nose simple, flat, bald, and with a central groove beneath.—Gray, s. *Viverrida.*

      1. Digitigrade: the under-side of the hind feet hairy, except the pads, metatarsus, and sometimes a small part of the tarsus. Upper flesh-tooth elongate; upper tubercular grinders small, transverse.—Gray.

         a. Body robust; tubercular grinders two above, one below on each side ($\frac{2-2}{1-1}$).—Gray.

VIVERRINAE. (A.)

b. Body slender, elongate; tubercular grinders one on each side above and below ($\frac{1}{1}$–$\frac{1}{1}$).—Gray.

PRIONODONTINAE. (B.)

2. Subplantigrade: the under-side of the toes and more or less of the back of the tarsus naked, callous. Flesh-tooth strong; upper tubercular grinders large, broad.

    a. Tail moderate, not prehensile. The hinder part of the tarsus hairy to the palm; the tail bushy.

GALIDIINAE. (C.)

    b. Tail moderate, not prehensile. The upper part of the hinder part of the tarsus hairy to the palm; tail ringed. (Gray.) Sectorial tooth with large tubercular ledge.

HEMIGALIINAE. (D.)

    c. Tail very long, sub-convolute. The hinder part of the tarsus bald, callous. (Gray.) Sectorial tooth typical.

PARADOXURINAE. (E.)

    d. Tail thick, strong, prehensile. The hinder part of the tarsus bald, callous. Sectorial tooth of upper jaw transverse, sub-tubercular.

ARCTICTIDINAE. (F.)

B. Nose rather produced, rounded, hairy, and without any central groove below (Gray, s. *Cynogalidæ*). Sectorial tooth with an extensive tubercular ledge.

CYNOGALINAE. (G.)

II. Auditory bulla very prominent and somewhat pyriform, divided by a transverse constriction into two portions; the anterior nearly as large and inflated as the posterior. (Flower.) Toes straight; the last phalanx and claws extended. The claws blunt and worn at the end, the front ones often elongated. (Gray, s. *Cynopoda*.)

A. Nose flat and bald, beneath with a central groove. (Gray, s. *Herpestidæ*.)

1. Head elongate, conical; tail conical or cylindrical. (Gray.)

HERPESTINAE. (H.)

2. Head short, ventricose; tail bushy, expanded laterally; claws elongate. (Gray.)

CYNICTIDINAE. (I.)

B. Nose broad, convex, and hairy, beneath without any central groove. (Gray, s. *Rhinogalidae*.)

1. Head elongate, nose short. Teeth 40. False grinders $\frac{3}{3}$. (Gray.)

RHINOGALINAE. (J.)

2. Head ventricose. Nose elongate. Teeth 36. False grinders $\frac{3}{3}$. (Gray.)

CROSSARCHINAE. (K.)

62

A. VIVERRINAE.

?. 1.

Viverra Linn.
Viverricula Hodgson.

§. 2.

Genetta Cuv.
  Genetta Gray.   ? *Fossa* Gray.

B. PRIONODONTINAE.

Prionodon Horsfield.=*Linsang* Gray.
Poiana Gray.

C. GALIDIINAE.

Galidia I. Geoff.

D. HEMIGALIINAE.

Hemigale Jourdan.

E. PARADOXURINAE.

Nandinia Gray.
Paradoxurus F. Cuv.
Paguma Gray.
Arctogale Peters.

F. ARCTICTIDINAE.

Arctictis Temm.=*Ictides* F. Cuv.

G. CYNOGALINAE.

Cynogale Gray.

H. HERPESTINAE.

Galidictis I. Geoff.
Herpestes Illig.
Athylax F. Cuv.
Calogale Gray.
Galerella Gray.
Calictis Gray.
Ariela Gray.
Ichneumia I. Geoff.
Bdeogale Peters.
Urva Hodgson.
Tœniogale Gray.
Onychogale Gray.
Helogale Gray.

I. CYNICTIDINAE.

Cynictis Ogilby.

J. RHINOGALINAE.

Rhinogale Gray.
Mungos Ogilby.

K. CROSSARCHINAE.

Crossarchus F. Cuv.
Suricata Desm.=*Rhyzæna*, Illig.

Extinct Viverridae?

Paleonyctis Blainv.
Soricictis Pomel.
Amphichneumon Pomel.
Galeotherium Waguer, (not Jacq.)

XIV. EUPLERIDAE.

Single genus.

Eupleres Doyère.

SUPER-FAMILY CYNOIDEA.

XV. CANIDAE.

SUB-FAMILIES.

I. Sectorial tooth of upper jaw elongated, and with the antero-internal lobe projecting directly inwards; of lower jaw, elongated and narrowed forwards and with the externo-median lobe enlarged: true molars in upper jaw two (rarely one), tubercular.

CANINAE. (A.)

II. Sectorial tooth of upper jaw abbreviated, triangular, and with the antero-internal lobe large and ledge-like; of lower jaw, comparatively short and broad forwards, and with the externo-median lobe reduced; true molars of upper jaw three, tubercular.

MEGALOTINAE. (B.)

A. CANINAE.

§. 1.

Lycaon Brookes.

§. 2.

Icticyon Lund=*Cynalicus* Gray=*Melictes* Schinz.

§. 3.

Cyon Hodgson.

§. 4.

Canis Linn.
  Canis=*Canis*+*Lupus* Gray.   *Dieba* Gray.
  Simenia Gray.   *Crysocyon* H. Smith.
Lycalopex Burm.
Pseudalopex Burm.
  Lycalopex Gray.   *Thous* Gray.

§. 5.

Vulpes.
  Vulpes.   *Leucocyon* Gray.
  Fennecus Gray.

§. 6.

Urocyon Baird.

Nyctereutes Temminck.

§. 7.

B. MEGALOTINAE.

Megalotis Blainv.=*Agriodus* H. Smith=*Otocyon* Licht.

Extinct Canidae? incertae sedis.

Amphicyon Lartet.
Cynodon Aym.
Galecynus Owen.
Palæocyon Lund, (not Blainv.)
Speothos Lund.

SUPER-FAMILY ARCTOIDEA.

XVI. MUSTELIDAE.

SUB-FAMILIES.

I. Skull with the cerebral portion comparatively compressed backwards; and with the rostral portion comparatively produced, attenuated, and transversely convex above; anteorbital foramen small and opening forwards. Feet with little developed or no interdigital membrane.

  A. Auditory bulla much inflated, undivided, bulging, and convex forwards; periotic region extending little outwards or backwards. Palate moderately emarginated.

    1. Last molar of upper jaw $\left(\text{M} \frac{1}{\cdot} \right)$ transverse, (with the inner ledge inflated at its inner angle;) sectorial tooth with a single inner cusp.

      a. M $\frac{1}{2}$; First true molar (sectorial) of lower jaw followed by a small second (tubercular) one. Toes short, regularly arched, and with the last phalanges bent up, withdrawing the claws into sheaths. (Gray.)

MUSTELINAE. (A.)

      b. M $\frac{1}{1}$; first true molar (sectorial) of lower jaw only developed. Toes straight, with the last phalanges and claws extended; the latter non-retractile. (Gray.)

MELLIVORINAE. (C.)

    2. Last molar of upper jaw $\left(\text{M} \frac{1}{\cdot} \right)$ enlarged and more or less extended longitudinally.—M $\frac{1}{2}$. Toes straight, with the last phalanges and claws extended; the latter non-retractile. (Gray.)

MELINAE. (B.)

  B. Auditory bulla elongated and extending backwards close to the paroccipital process. (Flower.) Palate moderately emarginated.

    1. Last molar of upper jaw $\left(\text{M} \frac{1}{\cdot} \right)$ transverse; (with the inner ledge narrowed inwards): sectorial tooth with two inner cusps.

HELICTIDINAE. (F.)

  C. Auditory bulla inflated, undivided, with the anterior inferior extremity pointed and commonly united to the prolonged hamular process of the pterygoid. (Flower.) Palate moderately emarginated.

1. Last molar of upper jaw $\left(M \frac{1}{\cdot}\right)$ transverse; (with the inner ledge compressed).

ZORILLINAE. (E.)

D. Auditory bulla little inflated, transversely constricted behind the meatus auditorius externus and thence inwards; in front flattened forwards: periotic region expanded outwards and backwards. Palate deeply emarginated.

1. Last molar of upper jaw $\left(M \frac{1}{\cdot}\right)$ quadrangular, wide, but with an extended outer incisorial ledge.

MEPHITINAE. (D.)

II. Skull with the cerebral portion swollen backwards and outwards; and with the rostral portion abbreviated, high and truncated forwards, and widened and depressed above: anteorbital foramen enlarged and produced downwards and backwards. Feet with well-developed interdigital membrane, and adapted for swimming.

A. Teeth normal, 36 (M $\frac{1}{2}$, PM $\frac{4}{3}$, C $\frac{1}{1}$, I $\frac{3}{3} \times 2$): sectorial tooth $\left(PM \frac{4}{\cdot}\right)$ normal, efficient, with an expanded inner ledge; the other molars submusteline. Posterior feet with normally long digits.

LUTRINAE. (G.)

B. Teeth very aberrant, 32 (M $\frac{1}{2}$, PM $\frac{3}{3}$, C $\frac{1}{1}$, I $\frac{3}{3}$—the lower inner incisors being lost—$\times 2$): sectorial tooth $\left(PM \frac{4}{\cdot}\right)$ defunctionalized as such, compressed from before backwards; the other molars also with blunted cusps. Posterior feet with elongated digits.

ENHYDRINAE. (E.)

A. MUSTELINAE.

§. 1.

(Digitigrade.)

Mustela L., Cuv.=*Martes* Gray (Les Martes—*Mustela* Cuv).
Putorius Cuv.=*Foetorius* Keys. and Blas.
  Putorius Gray.
  Gymnopus Gray.
  Gale Wagner=*Mustela* Gray, not Cuv.
  Lutreola Wagner=*Vison* Gray.

§. 2.

(Plantigrade.)

Galictis Bell=*Eirara* Lund.
  Galera Gray.   *Grisonia* Gray.

§. 3.

(Sub-plantigrade.)

Gulo Storr.

B. MELINAE.

Taxidea Waterh.
Meles Storr=*Taxus* Cuv.
Mydaus F. Cuv.
Arctonyx F. Cuv.=*Synarchus* Gloger.

August, 1871.

C. MELLIVORINAE.

Mellivora Storr=*Ratelus* Gray=*Lipo'us* Lund.

D. MEPHITINAE.

Conepatus Gray=*Thiosmus* Licht. < *Marpu'ius* Gray.
Mephitis Cuv., Gray.
Spilogale Gray.

E. ZORILLINAE.

Zorilla Gray=*Rhabdozale* Wagn.=*Ictonyx* Lund.

F. HELICTIDINAE.

Helictis Gray=*Melogale* I. Geoff.=*Rhinogale* Gloger, not Gray.

G. LUTRINAE.
§. 1.

Barangia Gray=*Leptonyx* Less., Gerv.
Aonyx Less., Gerv., Gray.
Lontra Gray=*Saricovia* Less.=*Loutra* Gerv. (misprint).
Lutra Linn.
   Lutra Gray.   *Nutria* Gray.
   Lutronectes Gray.
Hydrogale Gray.
Lutax Gray (not Gloger)=*Lataxia* Gerv.

§. 2.

Pteronura Gray, Gerv.=*Pterura* Wiegm.

H. ENHYDRINAE.

Enhydris Fleming=*Latax* Gloger.

Extinct Mustelidae? incertæ sedis.

Palaeomephitis Jäger=*Palaeobassaris* Paul von Wurt.
Palaeogale Meyer.
Plesiogale Pomel.
Plesictis Pomel.
Putoriodus Pomel.
Potamotherium Geoff.=? *Lutrictis* Pomel=? *Stephanodon* Meyer.
Thalassictis Nordm.
Galeotherium Jäger (not Wagner).
Enhydriodon Falc.=*Amyxodon* Falc.
Ursitaxus Falc.

XVII. URSIDAE.
Genera.
§. 1.

Thalassarctos Gray.
Ursus Linn.

Ursus Gray.   *Myrmarctos* Gray.
Tremarctos Gerv.
Helarctos Horsf.

§. 2.

Melursus Meyer=*Prochilus* Ill.

Extinct *Ursidae?*

(*Family? Hyænarctidae?*)

Hyænarctos Cautl. and Falc.=*Agriotherium* Wagn.=*Sivalarctos*+*Amphiarctos* Blainv.
=*Hemicyon* Lartet.

XVIII. AELURIDAE.

Genus.

Aelurus F. Cuv.

XIX. CERCOLEPTIDAE.

Genus.

Cercoleptes Illiger=*Kinkojou* Lac.=*Potos* Cuv.=*Caudivolvulus* Desm.

XX. PROCYONIDAE.

SUB-FAMILIES.

I. Snout attenuated. Auditory bulla small, abruptly contracted, flattened
forwards and towards the external auditory meatus. Mastoid process little
developed, extrorse behind meatus.

NASUINAE. (A.)

II. Snout comparatively abbreviated. Auditory bulla large, sloping gradually towards the external auditory meatus. Mastoid process enlarged and
prolonged downwards.

PROCYONINAE. (B.)

A. NASUINAE.

Nasua Storr=*Coati* Lac.

B. PROCYONINAE.

Procyon Storr.

Procyonidae? of extinct genera.

Tylodon Gerv.
Leptarctos Leidy.

XXI. BASSARIDIDAE.

Genus.

Bassaris Licht.

FISSIPEDIA INCERTAE SEDIS.

XXII. SIMOCYONIDAE.

Extinct.

Simocyon Kaup=*Diaphorus* Gaudry.

XXIII. ARCTOCYONIDAE.

Extinct.

Arctocyon Blainv.+*Palaeocyon* Blainv. (not Lund).

XXIV. HYAENODONTIDAE.

Extinct.

Hyaenodon de Laiz & de Par.=? *Hyaenodon*+*Taxotherium*+*Pterodon* Blainv.

Fissipedium Genera incertae sedis.

Acanthodon Meyer.
Harpagodon Meyer.
Patriofelis Leidy.
Sinopa Leidy.

(*Hyaenidae?*)

Lycyaena Hensel.
Hyaenictis Gaudry.

(*Viverridae?*)

Ictitherium Gaudry.

PINNIPEDIA.

FAMILIES.

I. Molar teeth $\frac{5}{5}$ or $\frac{6}{5}$: canines of both jaws moderately developed, those of upper jaw being scarcely larger than those of lower; incisors persistent. (*Phocoidea.*)

A. Form comparatively archetypical, with the hinder legs flexible forwards. Small ear conchs developed. Skull with the mastoid processes strong and salient, standing aloof from the auditory bullae; with well-developed post-orbital processes, and alisphenoid canals. Incisors ($\frac{6}{4}$) of upper jaw notched. Anterior limbs about as large as the posterior; their feet with digits decreasing in a curved line and without claws: posterior feet with all their digits nearly co-terminal and furnished with long linguiform flaps extending beyond their tips ; the three middle toes alone clawed.

OTARIIDAE. (XXV.)

B. Form attypically phociform, with the hinder legs projected backwards and not flexible forwards. Ear conchs obsolete. Skull with the mastoid processes swollen, and seeming to form part of the auditory bullae; the post-orbital processes null or obsolete; no alisphenoid canals. Incisors (variable in number—$\frac{6}{4}$ or $\frac{4}{4}$, or $\frac{4}{2}$—) of upper jaw not notched. Anterior limbs smaller than the posterior ; the feet with the digits successively abbreviated and armed with claws: the posterior flippers emarginated (the third and fourth digits being shortest), and provided with claws (rarely suppressed).

PHOCIDAE. (XXVI.)

II. Molar teeth $\frac{5}{4}$ $\frac{5}{4}$, the posterior generally caducous in adult: canines of upper jaw greatly hypertrophied and developed as tusks; those of lower jaw atrophied: incisors, except external of upper jaw, deciduous. (*Rosmaroidea.*)

A. Form comparatively etypical, with the hinder legs flexible forwards. Ear conchs obsolete. Skull with the mastoid processes strong and salient; the surface continuous with the auditory bullae; no postorbital processes; distinct alisphenoid canals. Anterior limbs about as large as posterior; feet with the toes decreasing in a curved line, destitute of claws: posterior feet with the five digits scarcely increasing toward inner; all provided with claws.

ROSMARIDAE. (XXVII.)

SUPER-FAMILY PHOCOIDEA.

XXV. OTARIIDAE.

Genera.

§. 1.

Zalophus Gill.
   Zalophus sensu strict.   *Neophoca* Gray.

§. 2.

Eumetopias Gill.
Otaria Peron.
   Otaria sensu strict.   *Phocarctos* Peters, Gray.
Arctocephalus F. Cuv.=*Halarctos* Gill.
   Arctocephalus Gray.
   Gypsophoca Gray.
   Arctophoca Peters=*Euotaria* Gray.
Callirhinus Gray.

XXVI. PHOCIDAE.

SUB-FAMILIES.

I. Maxillar zygomatic process with the posterior surface subvertical or very oblique. Malar oblong–rhomboid, emarginated above and below.

A. Intermaxillaries narrow, prolonged, and wedged behind between the supramaxillaries and nasals. Nasal bones narrow, diminishing in width backwards. Incisors $\frac{6}{4}$; exceptionally $\frac{4}{4}$.

PHOCINAE. (A.)

B. Intermaxillaries terminating far from nasals. Nasal bones narrow and shortened. Incisors $\frac{4}{2}$.

CYSTOPHORINAE. (B.)

II. Maxillar zygomatic process with its lower and posterior surface extended horizontally backwards, and its angle continued far behind along the inner side of the malar. Malar elongated, bow-shaped, and curved upward in front.

A. Intermaxillaries narrow, not continued backward between nasals and supramaxillaries. Nasal cavity expanded, with the nasal bones widest toward the middle and very long. Incisors $\frac{4}{4}$.

STENORHYNCHINAE. (C.)

A. PHOCINAE.

§. 1.

Phoca Linn., Gill=*Callocephalus* F. Cuv., Gray.
  Callocephalus Gray.   *Halicyon* Gray.
Pagomys Gray.
Fayophilus Gray.
Erignathus Gill=*Phoca* Gray, not Linn.

§. 2.

Halichoerus Nilss.

§. 3.

Monachus Flem.=*Pelagios* F. Cuv.=*Heliophoca* Gray.

B. CYSTOPHORINAE.

Cystophora Nilss.=*Stemmatopus* F. Cuv.
Macrorhinus F. Cuv.=*Mirounga* Gray=*Macrorhyna* Gray=*Morunga* Gray.

C. STENORHYNCHINAE.

Lobodon Gray.
Stenorhynchus F. Cuv.
Leptonychotes=*Leptonyx* Gray, not Sw. 1821.
Ommatophoca Gray.

Extinct Phocidae?

Pachyodon Meyer.
Pristiphoca Gerv.

SUPER-FAMILY ROSMAROIDEA.

XXVII. ROSMARIDAE.

Single genus.

Rosmarus Scop.=*Odobaenus* (Briss.) Ill.=*Trichechus* auct. pl., not Linn.

Extinct Rosmaridae.

Trichechodon Lankester.

UNGULATA.

SUB-ORDERS.

I. Digits paired, the third and fourth being subequally developed and exserted; (the fifth, generally, nearly corresponding in size and position to the second, and, generally, developed—or atrophied—in nearly equal degree;) the articulating phalanges and proximal carpal and tarsal bones correspondingly modified. Astragalus with its anterior or inferior articular surface divided by a crest into two sub-equal facets. Femur without a third trochanter, and with its shaft generally perforated at the fore and upper part by the medullary artery. Dorso-lumbar vertebrae, generally, nineteen in number (d. 12—15+1. 7—4.) Skull with the intermaxillary bones flattened above towards the symphysis, and with the incisors, when present, diverging towards their roots. Stomach more or less subdivided or complex: coecum comparatively small and simple.

ARTIODACTYLA.

II. Digits unpaired or unequal, the third being the largest and most exserted ; (the fourth nearly co-equal in size and position with the second ; fifth— of hind foot, at least,—atrophied ;) the articulating phalanges and carpal and tarsal bones correspondingly modified. Astragalus with the anterior or inferior articular surface divided into two very unequal facets. Femur with a third trochanter, and with its shaft perforated at the back-part by the medullary artery. Dorso-lumbar vertebrae not less than twenty-two in number (d. 18—19+1. 3—6). Skull with the intermaxillary bones tectiform above and united towards the symphysis, and with the incisors, when present, implanted subvertically and nearly parallel to their roots. Stomach simple: coecum very much enlarged and sacculated.

PERISSODACTYLA.

ARTIODACTYLA.

FAMILIES.

I. Molars (M) attypically each with two double crescentiform folds, whose convex surfaces are internal. Canines of lower jaw, attypically, resembling, and parallel with, incisors ; (differentiated and specialized in Camelidae). Palatine bones contracted and compressed behind, thin, and (at the walls of the posterior nares) separated by a wide sinus from the terminal portion of the supramaxillary bones. Digestive system adapted for rumination : stomach tripartite, or, attypically, quadripartite, a " psalterium " being finally developed.—Axis with the odontoid process like a spout, or hollow half-cylinder, and with a prominent sharp semi-circular rim. (Flower.)—(*Pecora ;* or, *Ruminantia.*)

* Incisors deciduous from upper as well as lower jaws. Canines of lower jaw inclined forwards, with compressed cuneate crowns. Placenta and stomach unknown. *Chalicotheroidea.*

One family. **CHALICOTHERIIDAE. (XXVII a.)**

** Incisors persistent in lower jaw.

A. Hind limbs with the proximal joint (femur) exserted and not contained within the common integument. Canines of lower jaw specialized and differentiated from incisors. Incisors in part (*i. e.* lateral) persistent in upper jaw. Placenta diffuse. Stomach imperfectly quadripartite. (*Pecora tylopoda s. phalangigrada.*)

One family. **CAMELIDAE. (XXVIII.)**

B. Hind limbs with the proximal joint (femur) not exserted but inclosed within the common integument. Canines of lower jaw similar to and parallel with the incisors. Incisors deciduous from upper jaw; persistent in lower. Placenta and stomach diversiform. (*Pecora unguligrada.*)

1. Placenta polycotyledonary. Stomach quadripartite, a well-developed psalterium being differentiated. Incisorial series of lower jaw uninterrupted at the symphysis. (*Pecora unguligrada typica.*)

a. Neck very long and slender, the cervical vertebrae (3–7) being much elongated: the dorso-lumbar vertebrae comparatively abbreviated and declining backwards, the hinder limbs being shorter than, or as short as, the anterior. Horns developed as epiphyses of the frontals, and covered with an extension of the skin. (*Giraffoidea.*)

One family. **GIRAFFIDAE.** (XXIX.)

b. Neck comparatively more or less short, the cervical vertebrae (3—7) being normally developed: the dorso-lumbar vertebrae longer, and highest backwards, the hinder limbs being considerably longer than the anterior. Horns diversiform. (*Booidea.*)

i. Skull with the auditory bulla produced downwards, especially towards the inside, and applied behind to the paroccipital process. Styloid process deflected more or less forwards and enclosed in an oblique fold on the outer surface of the auditory bulla. Palatine axis declivous from the occipito-sphenoid axis. (*Booidea typica.*)

a. Horns persistent, (common to both sexes,) and developed as sheaths of true "horn" on osseous cores originating from the frontal bones. Styloid process partially enclosed in a more or less open canal.

a. 1. Olfactory organ extremely expanded and inflated above: nasal bones much abbreviated, arched upwards, and entirely separated from the supra-maxillaries as well as lachrymals, the frontals projecting between the latter and the nasals. Supra-maxillaries and inter-maxillaries reduced and attenuated forwards. Posterior nasal cavity with walls inflated outwards.

SAIGIIDAE. (XXX.)

a. 2. Olfactory organ normally developed: nasal bones elongated, straight or declining forwards, and connected by suture with the lachrymals, supra-maxillaries and sometimes with the inter-maxillaries. Supra-maxillaries and inter-maxillaries well-developed forwards.

BOVIDAE. (XXXI.)

b. Horns deciduous, peculiar to the rutting season, (in both sexes,) developed as pseudocorneous sheaths with agglutinated hairs on osseous cores originating from the frontal bones. Styloid process completely inclosed in a canal by the lateral extension of the base of the bony meatus auditorius.

ANTILOCAPRIDAE. (XXXII.)

ii. Skull with the auditory bulla little produced downwards and applied only to the inner surface of the paroccipital process. Styloid process directed downwards, interposed between the bulla and paroccipital process, and not inclosed in an oblique fold of the auditory bulla. Palatine axis nearly parallel with the occipito-sphenoid axis. (*Booidea cerviformia.*)

One family. **CERVIDAE. (XXXIII.)**

2. Placenta diffuse. Stomach tripartite, the psalterium being undeveloped. Incisorial series of lower jaw interrupted at symphysis, (the middle incisors very enlarged and expanded towards their crowns.) (*Pecora unguligrada traguloidea.*)

One family. **TRAGULIDAE. (XXXIV.)**

3. Familiae incertae sedis.

a. Skull broad behind, in front of the molars contracted forwards, with the facial portion produced downwards and abbreviated, and with the nasal bones abbreviated and longitudinally arched. Molars (M $\frac{3}{3}$, PM $\frac{3}{3}$,) broad; inner crescentic plates of enamel running zig-zag-wise in large sinuous flexures. Horns in two pairs.

SIVATHERIIDAE. (XXXV.)

b. Skull with the parietals and supraoccipital extended far backwards, and contracted forwards in front of the molars, with the facial portion normally produced. Molars (M $\frac{3}{3}$, PM $\frac{3}{3}$,) broad; inner crescentic plates of enamel describing a simple curve. Horns none, (in both sexes?)

HELLADOTHERIIDAE. (XXXVI.)

C. Hind limbs with the proximal joint (femur) not exserted, but inclosed within the common integument (*Inferential*). Canines of lower jaw similar to and parallel with the incisors. Incisors all (I 3-3) persistent in upper jaw. (M $\frac{3}{3}$, PM $\frac{4}{4}$, C $\frac{1}{1}$, I $\frac{3}{3} \times 2 = 44$.) Placenta diffuse (*Inferential*). Stomach tripartite, the psalterium being undeveloped (*Inferential*). (*Pecora dentata.*

1. Teeth of both jaws in an interrupted series, the canines of the upper jaw being enlarged, and the first premolar of the lower caniniform, and received in diastemas of the opposite jaw. (*Oreodontoidea.*)

OREODONTIDAE. (XXXVII.)

2. Teeth of both jaws in a nearly or quite uninterrupted series, the canines and first premolars of neither jaws projecting. (*Anoplotheroidea.*)

a. Body somewhat cerviform, with the hind limbs little longer than the fore, (having the relative length normal to walking quadrupeds.) Teeth comparatively uniform.

ANOPLOTHERIIDAE. (XXXVIII.)

b. Body somewhat leporiform, with the hind limbs much longer than the fore, (as in the Leporids.) Teeth comparatively differentiated.

DICHOBUNIDAE. (XXXIX.)

II. Molars (M) attypically tuberculiferous. Canines of lower jaw enlarged and often developed as tusks, entirely differentiated and distant from incisors. Palatine bones scarcely contracted behind, thick, and (at the walls of the posterior nares) articulated with the terminal portion of the supramaxillary bones. Digestive system not adapted for rumination: stomach imperfectly septate.—Axis with the odontoid process conical. (Flower.)—(*Omnivora.*)

A. Body massive, with the feet phalangigrade, and their external (2, 5) toes well developed and produced as far as or beyond the first phalanges of the middle (3—4) toes; the last phalanges wide and with convex margins: manus with unciform bone much broader than high, and with second phalanx wedged between trapezoid and magnum; pes with cuboid broader than high. Lower jaw with a deep preangular expansion directed forwards. (Snout rounded and with the nostrils open upwards and sideways. Mammae two, inguinal.) *Obesa.*

? Molars of upper jaw with a bow-shaped (convex extrorsely) longitudinal and a straight transverse valley dividing four tubercles, all of which are convex introrsely (towards the palate) and concave externally, (thus simulating the teeth of ruminants.) Molars of lower jaw narrower than those of upper, and with the longitudinal valley very narrow: (last molar with a supplementary posterior lobe.) Canines comparatively small and cylindro-conic. (*Merycopotamoidea.*)

MERYCOPOTAMIDAE. (XL.)

! Molars (M) of upper jaw with nearly straight or irregularly sinuous longitudinal and transverse valleys dividing four tubercles, of which the external two are convex extrorsely and the inner two convex introrsely (towards the palate.) Molars of lower jaw resembling those of upper, (the last molar with a supplementary posterior lobe.) Canines very large and furrowed along their posterior surface. (*Hippopotamoidea.*)

HIPPOPOTAMIDAE. (XLI)

B. Body suiform; with the feet unguligrade, and their external toes reduced in size and not produced or assisting in progression; the last phalanges elongated and trihedral: manus with the unciform little or no broader than deep, and with the second phalanx not wedged between the trapezoid and magnum; pes with cuboid deeper than broad and emarginated behind. Lower jaw with no preangular expansion. (Snout disciform and with the nostrils in it and open forwards. Mammae in increased number (4 to 10), ventral as well as inguinal.) *Setifera.*

1. True molars of upper jaw with oblong crowns with four or more principal sub-conical lobes and accessory smaller ones.

a. Occipital bone with long deflected styliform paroccipital processes in front of the occipital condyles, and emitting transverse internal ridges in which are the condyloid foramina. Squamosals with their articular processes projecting directly outwards from their bases (and thus aloof from the auditory bullae), and with the zygomatic processes overlying the malar bones. Pterygoid bones twisted and reflected outwards : the crest continued upwards and backwards into the temporal region. Articular surface for lower jaw transversely concave, antero-posteriorly convex, and with no post-glenoid process. Lower jaw with triangular condyles. Canine teeth of upper jaw (in males) more or less twisted outwards and upwards and parallel with the lower. Back with no dorsal scent gland. (*Setifera suiformia.*)

 i. Skull with the palato-maxillary axis extremely deflected and forming a high angle with the occipito-sphenoidal axis. Basisphenoid reflected (with a crest uniting with the presphenoid), and forming two deep pocket-like cavities. Orbits directed upwards and backwards. Malar bones very deep, and with a short inferior process. Dental series aberrant (molars reduced (in old) to true (M 1—3) or even last true molar) : last or third true molar elongated and composed of three longitudinal rows of columnar tubercles presenting, when worn, simple oval insular areas. (Incisors, in adults, reduced to 2 (or none) in upper, and sometimes none in lower jaw.)

PHACOCHOERIDAE. (XLII.)

 ii. Skull with the palato-maxillary axis little deflected, and nearly parallel with the occipito-sphenoidal axis. Basisphenoid normal, and with no bursiform cavities. Orbits directed outwards and forwards. Malar bones elongated and with a long inferior process. Dental series normal (M $\frac{3}{3}\times2$, PM $\frac{4}{4}\times2$, C $\frac{1}{1}\times2$, I $\frac{3}{3}\times2=44$): molars with corrugated cusps presenting, when worn, deeply sinuated insular areas.

SUIDAE. (XLIII.)

b. Occipital bone with short backward-directed paroccipital processes originating sideways from the occipital condyles, and emitting a transverse internal ridge continuous with the anterior margin of the bone, behind which are the condyloid foramina. Squamosals with their articular processes deflected from their bases and bounding the outside of the auditory bullae, and with the zygomatic processes articulating obliquely with the malar bones. Pterygoid bones simply curved outwards : the crest with a crest-like anterior process of the squamosal in front of the auditory bullae. Glenoid fossa curved and transversely concave, antero-posteriorly concave and with a distinct post-glenoid process. Lower jaw with transverse condyles. Canine teeth of upper jaw simply decurved, very acute and trenchant behind. Back with a posterior dorsal scent gland. (*Setifera dicotyliformia.*)

 One family.   DICOTYLIDAE. (XLIV.)

2. True molars of upper jaw with quadrate crowns, with four principal pyramidal and more or less distinctly trihedral lobes, divided by deep valleys, not filled up by cement, but, in some genera, interrupted with minor tubercles and ridges. (Owen.) Orbits, attypically, with a continuous margin behind. Lower jaw, attypically, with a tubercle projecting outwards. (*Anthracotheroidea.*)

ANTHRACOTHERIIDAE. (XLV.)

ARTIODACTYLI? INCERTAE SEDIS.

SUPER-FAMILY CHALICOTHEROIDEA.

XXVIIIa. CHALICOTHERIIDAE.

Chalicotherium Kaup., Falc.

PECORA.

SUPER-FAMILY CAMELOIDEA.

XXVIII. CAMELIDAE.

Genera.

Camelus Linn.
Auchenia Ill.

Extinct Camelidae.

Merycotherium Bojanus.
Poebrotherium Leidy.
Procamelus Leidy.
Megalomeryx Leidy.
Homocamelus Leidy.
Protomeryx Leidy.
Merycodus Leidy.
Camelops Leidy.
Pulauchenia Owen.

SUPER-FAMILY GIRAFFOIDEA.

XXIX. GIRAFFIDAE.

Single genus.

Giraffa Storr ex. Briss.=*Camelopardalis* Cuv.

SUPER-FAMILY BOOIDEA

XXX. SAIGIIDAE.

Genus.

Saiga Gray.

XXXI. BOVIDIAE.

SUB-FAMILIES.

(*Fide auct. plur.*)

I. Form massive, with the head declined; with the neck abbreviated, the third and succeeding vertebrae being much shortened. Legs stout, and with the metacarpals and metatarsals little or no longer than the phalanges with hoofs.

A. Molars comparatively broad, without supplemental lobes. The basi-occipital bone with its tubercles well developed, and a deep groove between them. (Turner.)

BOVINAE. (A.)

B. Molars comparatively narrow, with supplemental lobes. The basioccipital bone broad and flat, with a ridge and a fossa on each side. (Turner.)

OVIBOVINAE. (B.)

II. Form slender, with the head more or less uplifted; with the neck comparatively lengthened, the third and succeeding vertebrae being not much shorter than thick. Legs slender, and with the metacarpals and metatarsals much longer than the phalanges with hoofs.

1. Horns diversiform (definable by no common characters), conical, cylindrical, or compressed; or, sub-angular, with a sub-spiral ridge originating at the base anteriorly; or, variously contorted.

ANTILOPINAE. (C.)

2. Horns curved backwards, sub-angular, with a rectilinear ridge anteriorly continuous around the convex curve.

CAPRINAE. (D.)

3. Horns curved outwards and forwards or sub-spiral, sub-angular, with a rectilinear ridge continuous around the convex curve.

OVINAE. (E.)

A. BOVINAE.

Bos Linn.
Bibos Hodgson.
  Bibos sensu strict.
Bubalus H. Smith.
  Bubalus sensu strict.
Hemibos Falc. (*Extinct.*)
Anoa Leach.
Poëphagus Gray.
Bison H. Smith=*Bonasus* Wagn.

Probos Hodgson.

Syncerus Hodgson.
Amphibos Falc. (*Extinct.*)

B. OVIBOVINAE.

Oribos Blainv.
  Ovibos sensu strict.

Bootherium Leidy. (*Extinct.*)

C. ANTILOPINAE.

§. 1.

(*Strepsiceros* Turner.)

Strepsiceros H. Smith.
Oreas Desm.
Tragelaphus Blainv.

§. 2.

(*Gazella* Turner.)

Pantholops Hodgson, Gray, Gerv.
Procapra Hodgson.
Gazella Blainv.
Tragops Hodgson.
Antidorcas Sund.

(*Antilope* Turner.)

Æpyceros Sund.

(*Cervicapra* Turner.)

Antilope Blainv.

(*Tetraceros* Turner.)

Tetraceros Leach.

(*Oreotragus* Turner.)

Calotragus Sund.
Scopophorus Gray.
Oreotragus Gray.

(*Neotragus* Turner.)

Nesotragus Von Duben.

(*Cephalophus* Turner.)

Cephalophus H. Smith.

(*Eleotragus* Turner.)

Nanotragus Sund.
Pelea Gray.
Eleotragus Gray.
Adenota Gray.
Kobus H. Smith.

§. 3.

(*Catoblepas* Turner.)

Connochetes Licht.
  Connochetes Gray.   *Gorgon* Gray.

(*Alcelaphus* Turner.)

Alcelaphus Blainv.
Damalis H. Smith=*Gazella* §. Gerv.

§. 4.

(*Nemorhaedus* Turner.)

Capricornis Ogilby.
Nemorhaedus H. Smith.

(*Budorcas* Turner.)

Budorcas Hodgson.

§. 5.

(*Apolceros* Turner.)

Mazama Raf., Gray=*Aploceros* H. Smith=*Antilocapra* Gerv.

(*Rupicapra* Turner.)

Rupicapra Blainv., Gray=*Capella* K. and B.

§. 6.

Aegoceros H. Smith, Turner=*Hippotragus* Sund.
Oryx Blainv., Turner.
Addax Gray, Turner.

§. 7.

(*Portax* Turner.)

Portax H. Smith.

D. CAPRINAE.

Hemitragus Gray.
  Hemitragus Gray.
  Kemas Ogilby, Gray, Gerv.
Capra Linn.
  Aegoceros (Pall., Gray (p. 147, not p. 142).
  Ibex (Pall.), Gerv.=*Capra* Gray.
  Capra (Linn.), Gerv.=*Hircus* Gray.

E. OVINAE.

Ovis Linn.
  Ovis sensu strict.
  Caprovis Hodgson=*Musimon* Gray, Gerv.
Pseudovis Hodgson.
Ammotragus Blyth.

Extinct genera.

(*Antilopinae.*)

Palaeotragus Gaudry.
Palaeoryx Gaudry.
Tragoceros Gaudry.
Palaeoreas Gaudry.
Antidorcas Gaudry.

(*Bovidae? incertae sedis.*)

Leptotherium Lund.
Cosoryx Leidy.

XXXII. ANTILOCAPRIDAE.

Genus.

Antilocapra Ord=*Dicranoceros* H. Smith.

XXXIII. CERVIDAE.

SUB-FAMILIES.

I. Horns developed.
  A. Canine teeth small or none.
    CERVINAE. (A.)
  B. Canine tooth of male enlarged and tusk-like
    CERVULINAE. (B.)
II. Horns not developed.
  A. Canine teeth of male enlarged and tusk-like.
    MOSCHINAE. (C.)

A. CERVINAE.

(*Genera fide* Sclater.)

§. 1.

Alces H. Smith.

§. 2.

Rangifer H. Smith=*Tarandus* Ogilby.

§. 3.

Dama H. Smith.
Cervus Linn., Sclater.
  Cervus sensu strict.
  Elaphurus A. M. Edw.
  Rusa Hodgson.
  Axis Hodgson.
  Blastoceros Sund.
  Coassus Gray.
Capreolus Gray.

  Sika Hodgson.
  Rucervus Hodgson=*Panolia*
  Hyelaphus Sund.   [Gray.
  Cariacus Gray.
  Furcifer Sund.
  Pudu Gray.

B. CERVULINAE.

Cervulus Blainv.=*Muntjacus* Gray=*Stylocerus* H. Smith=*Prox* Ogilby.

C. MOSCHINAE.

Moschus Linn.
Hydropotes Swinhoe.

Extinct.

(*Cervinae.*)

Megaceros Owen.

(*Cervidae? related to Moschinae?*)

Dremotherium E. Geoff.
Amphitragulus Pomel=*Tragulotherium* Croizet.
Dorcatherium Kaup.
Leptomeryx Leidy.

SUPER-FAMILY TRAGULOIDEA.

XXXIV. TRAGULIDAE.

Genera.

§. 1.

Tragulus Briss.
  Tragulus sensu strict.   *Meminna* Gray.

§. 2.

Hyomoschus Gray.

SUPER-FAMILY? SIVATHEROIDEA.

XXXV. SIVATHERIIDAE.

Extinct.

Sivatherium Falc. and Cautl.

Incertæ sedis.

Bramatherium Falc. and Cautl.

SUPER-FAMILY? HELLADOTHEROIDEA.

XXXVI. HELLADOTHERIIDAE.

Extinct.

Helladotherium Gaudry.

SUPER-FAMILY OREODONTOIDEA.

XXXVII. OREODONTIDAE.

Extinct.

A. Orbit complete behind.   Lachrymal bone impressed by a well-marked fossa.   (Leidy.)

OREODONTINAE. (A.)

B   Orbit incomplete behind.   Lachrymal bone with no fossa.   (Leidy.)

AGRIOCHOERIDAE. (B.)

A. OREODONTINAE.

Oreodon Leidy=*Merycoidodon* Leidy=*Cotylops* Leidy.   (*Fide* Leidy.)
Merycochoerus Leidy.
Merychyus Leidy.
Leptauchenia Leidy.

B. AGRIOCHOERIDAE.

Agriochoerus Leidy.=? *Eucrotaphus* Leidy.

SUPER-FAMILY ANOPLOTHERIOIDEA.

XXXVIII. ANOPLOTHERIIDAE.

Extinct.

Anoplotherium Cuv.
Eurytherium Gervais.

XXXIX. DICHOBUNIDAE.

Extinct.

(*Genera fide* Turner.)

Caenotherium Bravard=*Oplotherium* Laiz. and de Par.
Dichodon Owen.
Dichobune Cuv.
Xiphodon Cuv.

Anoplotheroidea? incertæ sedis.

Tapinodon v. Meyer, 1846.
Choereomeryx Pomel, 1848.
Aphelotherium Gervais.

February, 1872.

Cebochoerus Gervais.
Zooligus Aymard.
Diplocus Aymard.
Hyaegulus Pomel.
Microtherium v. Meyer=*Amphimeryx* Pomel.
Adapis Cuv.
Homaladotherium Huxl.

OMNIVORA.

SUPER-FAMILY MERYCOPOTAMOIDEA.

XL. MERYCOPOTAMIDAE.

Extinct.

Merycopotamus Falc. and Cautl.

SUPER-FAMILY HIPPOPOTAMOIDEA.

XLI. HIPPOPOTAMIDAE.

SUB-FAMILIES.

A. Skull depressed between the orbits and with the frontal sinus obsolete; the orbits prominent above the level of the forehead, and closed behind.

HIPPOPOTAMINAE. (A.)

B. Skull convex between the orbits and with the frontal sinus well developed; the orbits depressed below the level of the forehead and incomplete behind.

CHOEROPSINAE. (B.)

A. HIPPOPOTAMINAE.

Hippopotamus Linn.=*Tetraprotodon* Falc. and Cautl.

B. CHOEROPSINAE.

Choeropsis Leidy.

Extinct.

(*Hippopotaminae.*)

Hexaprotodon Falc. and Cautl.

SUPER-FAMILY SETIFERA.

XLII. PHACOCHOERIDAE.

Genus.

Phacochoerus F. Cuv.=*Eureodon* G. Fisch.

Extinct genus referred (erroneously?) to Phacochoeridae.

Calydonius v. Meyer.

XLIII. SUIDAE.

§ 1.

Babirussa F. Cuv.=*Porcus* Wagler.

§. 2.

Potamochoerus Gray=*Choiropotamus* Gray.
Sus Linn.
  Sus Gray.   *Scrofa* Gray.
  Centuriosus Gray=*Gyrosus* Gray=*Ptychochoerus* Fitz.
Porcula Hodgson.

Extinct genus incertæ sedis.

Hippohyus Falc. and Cautl.

XLIV. DICOTYLIDAE.

Genera.

Dicotyles Cuv.
Notophorus Gray.

Extinct.

Platygonus Lec., Leidy.=*Hyops* Lec.=*Protochoerus* Lec.=*Euchoerus* Leidy. (*Fide* Leidy.)

SUPER-FAMILY ANTHRACOTHEROIDEA.

XLV. ANTHRACOTHERIIDAE.

Extinct.

SUB-FAMILIES.

A. Premolars of upper jaw in part (PM 4) resembling the true molars, and with tubercles in transverse series $\left(\frac{1}{1} \mid \frac{1}{1\cdot2} \right)$ separated by transverse vallies; the preceding (PM 3, 2, 1) successively more and more differentiated forwards.

                                         ' HYOPOTAMINAE. (A.)

B. Premolars (PM 4, 3, 2, 1) of upper jaw all differentiated from the true molars, and each with a conical crown and a small inner lobe.

                                      ANTHRACOTHERIINAE. (B.)

A. HYOPOTAMINAE.

Hyopotamus Owen.
Bothryodon Aymard=*Ancodus* Pomel.   •

B. ANTHRACOTHERIINAE.

Anthracotherium Cuv.
Elotherium Pomel.

EXTINCT OMNIVORA? INCERTÆ SEDIS.

Choeropotamus Cuv.
Palaeochoerus Pomel = *Cyclognathus* Croizet = *Brachygnathus* Pomel = *Synaphodus* Pomel.
Choeromorus Lartet.
Entelodon Aymard.
Heterohyus Gervais.
Acotherulum Gervais.
Chocrotherium Falc.=*Tetraconodon* Falc.

Titanotherium Leidy.
Perchoerus Leidy.
Leptochoerus Leidy.
Nanohyus Leidy.

PERISSODACTYLI.

FAMILIES.

I. Incisors (4? in lower jaw) in part gliriform, the outer having persistent pulps, and growing continuously in a circular direction. (*Anchippodonto-idea.*)

ANCHIPPODONTIDAE. (XLV. a.)

II. Incisors not gliriform.

1. Middle digit and hoof hypertrophied and alone supporting the body, the lateral (second and fourth) digits being more or less atrophied and functionless, or (attypically) obsolete (reduced to the condition of "splint bones"). Femur with a fossa above the external condyle. Skull (attypically) much prolonged forwards. Molars subequal (not decreasing forwards) and cuboid; pre-molars (PM 3-4) also enlarged (not decreasing forwards) and (except second) cuboid; the second (PM 2) elongated; the first milk molar (D 1) more or less persistent and not replaced by a pre-molar (PM 1); disproportionately small. Incisors with a deep invaginated fold of enamel penetrating the interior from the crown, and producing a central cavity filled with cementum. (*Solidungula.*)

A. Ulna with the shaft atrophied and the extremities anchylosed and consolidated with the radius. Fibula rudimentary and anchylosed to the tibia. Skull with the orbit complete behind. Upper molars (PM and M)—at least, of second set—with a deep valley re-entering from the postrorse portion of the inner side obliquely forwards, and (in connection with a more or less deep valley re-entering from the introrse portion of the anterior border or the angle) more or less isolating an introrse enamel lobe or area, and with two (anterior and posterior) crescentic enamel islands. Lower molars (PM 2, M 2) with a valley re-entering inwards from the outer wall, one from the introrse portion of the anterior wall, and another (terminating in anterior and posterior branches) from the posterior portion of the inner wall.

EQUIDAE. (XLVI.)

B. Ulna with the shaft complete and moderately developed, and more or less differentiated from the radius. Fibula archetypically complete but anchylosed with the tibia. Skull with the orbit incomplete behind. Upper molars (PM 3-4 and M) with a deep (anterior) valley re-entering from the middle of inner side inwards and forwards

and ending in lateral branches, and with a (posterior) valley re-
entering from the posterior wall. Lower molars with a V-shaped
valley re-entering from the outer wall, and two V-shaped vallies,
re-entering from the inner wall (the crowns having W-shaped
ridges)

ANCHITHERIIDAE. (XLVII.)

2. Middle digit and hoof not hypertrophied, and only in connection with
the lateral supporting the body, the lateral being well developed and
efficient. Femur without a fossa above the external condyle. Skull
moderately prolonged forwards. Molars unequal (the first smaller
than the second), diversiform; pre-molars decreasing in size forwards;
first milk molar not persistent, but (generally) replaced by a pre-
molar (PM 1) of moderate size. Incisors without an invaginated fold
of enamel penetrating the interior.

A. Nasal region expanded or thrown backwards, the supramaxillary
bones forming a more or less considerable portion of the border of
the nasal aperture; the nasal bones contracted forwards, or atrophied. Molars with crowns traversed by more or less well-defined
continuous ridges.

   a. Upper molars with a continuous outer wall and without complete transverse crests. (*Rhinocerotoidea.*)

      aa. Neck abbreviated. Incisor teeth (attypically) reduced in
      number or entirely suppressed. (*Rhinocerotoidea rhinoceroti-*
      formia.)

         * Skull with the basioccipital comparatively well developed
         behind and narrowed forwards; (with tympanic and periotic
         bones anchylosed and wedged in between the squamosal, ex-
         occipital and other adjacent cranial bones.—Huxley;) with
         the nasal bones produced forwards and more or less arched,
         and meeting an upward developed expansion of the supra-
         maxillary bones. Upper molars with a deep valley extending
         obliquely inwards from the median portion of the inner wall
         and (PM 4, M 1–2) a shallow one extending from the posterior
         wall. Lower molars (PM 2, M 3) with two curved transverse
         crests.

RHINOCEROTIDAE. (XLVIII.)

      bb. Neck more or less elongated. Incisor teeth developed in
      normal number ($\frac{6}{6}$). (*Rhinocerotoidea macraucheniformia.*)

         * Skull with the basioccipital widened forwards: with the nasal
         bones extremely reduced and above or behind the orbits: the
         supramaxillary bones rectilinear above, arched and approxi-
         mating each other in front of the nasal aperture but separated
         by the extension upward of the vomer? Dental series almost

continuous: upper posterior molars (M 2, 3) each with a shallow valley extending inwards from the anterior portion of the inner wall, and with two or three deep depressions in the inner half: lower molars (PM 3, M 3) with two (anterior and posterior) more or less defined crescent-shaped ridges: canines small.

MACRAUCHENIIDAE. XLIX.)

** Skull with the basioccipital comparatively narrow forwards: with the nasal bones produced forwards and terminating in a free narrowed surface; the supra-maxillary bones with an upward developed expansion (connected with the nasal bones) and widely separated above in front. Dental series interrupted by wide diastemas: upper molars (PM 2, M 1, 2, 3) each with a deep valley extending obliquely inwards from the median portion of the inner wall and a shallow one extending from the angle or posterior wall: lower molars (PM 2, M 2) with two (anterior and posterior) crescent-shaped ridges: canines well developed.

PALAEOTHERIIDAE. (L.)

b. Upper (as well as lower) true molars without a continuous outer wall, but (M 2–3, at least,) each with two complete transverse crests. (*Lophiodontoidea.*)

1. True molars as well as pre-molars in part (PM 2, 3, 4) nearly similar, squarish, and each with the anterior crest marginal, but with an anterior cingulum terminating in a cusp at the antero-outer angle of the tooth; hindmost molar (M 3) with no posterior lobe. Anterior feet with four toes; posterior with three, (in known types).

TAPIRIDAE. (LI.)

2. True molars and pre-molars differentiated from each other; the former oblong, with the anterior crest remote from the anterior margin and continuous with a small crest recurrent from the outer wall: hindmost molar with a posterior lobe; (pre-molars not bilophodont but with a lobe extending inwards from the inner wall). Anterior feet with four (?) toes; posterior with three (?).

LOPHIODONTIDAE. (LII.)

B. Nasal region compressed and extended forwards, the supramaxillary bones excluded from the nasal aperture; the nasal bones elongated and extending far forwards, and articulated with the intermaxillary bones. Molars (M 1, 2, 3) of upper jaw each with two transverse rows of tubercles (3|3) separated by a transverse valley and with a cingulum anteriorly and internally: (lower molars dissimilar). (*Pliolophoidea.*)

PLIOLOPHIDAE. (LIII.)

Perissodactyli? incertae sedis.

Molar teeth of lower jaw with a crenulated longitudinal ridge. Canines and incisors wanting.

ELASMOTHERIIDAE. (LIV.)

SUPER-FAMILY ANCHIPPODONTOIDEA.

XLV a. ANCHIPPODONTIDAE.

Extinct.

Anchippodus Leidy=*Trogosus* Leidy.

SUPER-FAMILY SOLIDUNGULA.

XLVI. EQUIDAE.

Genera.

Equus Linn.
Asinus Gray.
   Asinus sensu strict.   *Hippotigris* H. Smith.

Extinct.

Hipparion Christol=*Hippotherium* Kaup.
Merychippus Leidy.
Protophippus Leidy=*Hippidion* Owen 1870.

XLVII. ANCHITHERIIDAE.

Extinct.

Genera fide Leidy.

Anchitherium v. Meyer=*Hipparitherium* Christol.
Hypohippus Leidy, 1858.
Parahippus Leidy, 1858.
Anchippus Leidy, 1868.

SUPER-FAMILY RHINOCEROTOIDEA.

XLVIII. RHINOCEROTIDAE.

Genera.

Rhinaster Gray.
   Rhinaster sensu strict.   *Ceratotherium* Gray.
Rhinoceros Linn.
   Rhinoceros sensu strict.   *Ceratorhinus* Gray.

Extinct.

§. 1.

Coelodonta Bronn.

§. 2.

Acerothirium Kaup.
Badactherium Croizet.
Hyracodon Leidy.

XLIX. MACRAUCHENIIDAE.

Extinct.

Macrauchenia Owen=*Opisthorhinus* Bravard.

L. PALAEOTHERIIDAE.

Extinct.

Palaeotherium Cuv.
Monacrum Aymard.
Propalaeotherium Gervais, 1849.
Paloplotherium Owen=*Plagiolophus* Pomel.

SUPER-FAMILY LOPHIODONTOIDEA.

LI. TAPIRIDAE.

Genera.

§. 1.

Elasmognathus Gill.

§. 2.

Tapirus Linn.
Rhinochoerus Gray.

LII. LOPHIODONTIDAE.

Extinct.

Genera fide Bronn.

Lophiodon Cuv.=*Tapirotherium* Blainv. 1817 (not 1846).
Tapiroporcus Jäger, 1835.
Coryphodon Owen, 1846.
Listriodon v. Meyer, 1846=*Tapirotherium* Lartet
Pachynolophus Pomel, 1847=*Hyracotherium* Blainv. 1844 (not Owen, 1840).
Lophiotherium Gervais, 1849.
Tapirulus Gervais, 1850.
Anchilophus Gervais, 1852.

SUPER-FAMILY PLIOLOPHOIDEA.

LIII. PLIOLOPHIDAE.

Extinct

Pliolophus Owen.

PERISSODACTYLI INCERTAE SEDIS.

LIV. ELASMOTHERIIDAE.

Extinct.

Elasmotherium Fischer =? *Stereoceros* Duvernoy.

UNGULATA? INCERTAE SEDIS.

Hyracotherium Owen.
Stereognathus Owen.

IV. TOXODONTIA.

FAMILIES.

I. Teeth 44 (M $\frac{3}{3}$, PM $\frac{4}{4}$, C $\frac{1}{1}$, I $\frac{3}{3}\times2$) ; molars of upper jaw mostly (PM 3–4, M 1) oblong, moderately narrowed backwards, with two folds (the anterior of which is divided) re-entering from the inner side. Incisors three on each side, with simple fangs ; the first largest, compressed, widely separated from its homologue; the second smaller, trihedral; the third lateral and behind the second, and rudimentary: molars of lower jaw comparatively broad and complex : canines moderate ; incisors implanted in a curved row.

NESODONTIDAE. (LV.)

II. Teeth 36 (M $\frac{3}{3}$, PM $\frac{4}{4}$, C $\frac{0}{0}$, I $\frac{2}{3}\times2$) ; molars with enamel coat interrupted at the anterior and posterior margins ; those of upper jaw mostly (PM 3, 4, M 1–3) obliquely triangular, rapidly narrowed backwards, with a single simple fold re-entering obliquely forwards from the inner side. Incisors of upper jaw two on each side, but with incisorial crowns, the outer with roots of nearly uniform diameter throughout, and describing the segment of a circle, (like those of rodents,) and with persistent pulp—(Owen): molars of lower jaw narrowed, especially the posterior portions ; canines rudimentary ; incisors in a nearly straight line.

TOXODONTIDAE. (LVI.)

LV. NESODONTIDAE.

Extinct.

Nesodon Owen.

LVI. TOXODONTIDAE.

Extinct.

Toxodon Owen

V. HYRACOIDEA.

FAMILY.

LVII. HYRACIDAE.

Genera.

Hyrax Linn.
  Hyrax Gray.
Dendrohyrax Gray.

Euhyrax Gray.

IV. PROBOSCIDEA.

FAMILIES.

I. Incisors of upper jaw (1+1) everted, enormously developed and modified as cylindro-conic tusks, with roots extending backwards and converging, and thus producing a high pre-narial rampart: incisors of lower jaw comparatively small and directed forwards, or entirely absent. Molars successively displacing each other from behind forwards (and therefore no pre-molars replacing the deciduous ones), and not more than two (or one) fully developed at the same time. Skull abbreviated and enlarged obliquely, convex backwards and outwards, and with the occipital condyles declined.

ELEPHANTIDAE. (LVIII.)

II. Incisors of upper jaw atrophied or absent, (and consequently an uninterrupted oval depression occupying the naso-maxillary region): incisors of lower jaw (1+1) enlarged, and developed as tusks decurved downwards and backwards, and involving the symphysial portion of the jaw. Molars vertically developed (with pre-molars replacing the deciduous molars), and in considerable number (PM $\frac{2}{2}$, M $\frac{3}{3}\times2$) at the same time. Skull moderately long, and with the occipital condyles inclined upwards.

DINOTHERIIDAE. (LIX.)

LVIII. ELEPHANTIDAE.

SUB-FAMILIES.

I. Intermediate molars (D 4, M 1, 2) with an "isomerous" ridge formula (i. e. with the ridges equal in number in the successive teeth—three to five): the ridges attypically continuous: the valleys with a thick deposit of cementum.

ELEPHANTINAE. (A.)

II. Intermediate molars (D 4, M 1, 2) with a "hypisomerous" or "anisomerous" ridge formula (i. e. with the ridges increasing in number by one ("hypisomerous") or more ("anisomerous") in the successive teeth (e. g. D 4⁷, M 1⁸, M 2⁹ to D 4¹² ᵖᵐ·, M 1¹⁶ ᵖᵐ·, M 2¹⁸ ᵖᵐ·): the ridges with more or less mammilliform tubercles: the valleys with little or no cementum.

MASTODONTINAE. (B.)

A. ELEPHANTINAE.

Elephas Linn=*Elasmodon* Falc.=*Euelephas* Falc.
Loxodonta F. Cuv.=*Loxodon* Falc.

Extinct genus.

Stegodon Falc.

B. MASTODONTINAE.

Extinct.

Pentalophodon Falc.
Mastodon Cuv.=*Tetralophodon* Falc.
Tetracaulodon Godman=*Trilophodon* Falc.

LIX. DINOTHERIIDAE.

Extinct.

Dinotherium Kaup.

VII. SIRENIA.

FAMILIES.

I. Tail entire, rounded, and with the vertebrae towards last (*i. e.* $5+x$); sub-cylindrical and destitute of transverse processes. Intermaxillary bones with the branches little prolonged backwards and with the anterior portion nearly or quite straight. (*Trichechoidea.*)

TRICHECHIDAE. (LX.)

II. Tail forked, and with the vertebrae (except the terminal) depressed and provided with transverse processes. Intermaxillary bones (attypically) with the branches prolonged backwards and with the anterior portion more or less deflected. (*Halicoroidea.*)

A. Teeth present, and in part at least functionally developed in the adult: molars $\frac{5}{5}$ to $\frac{6}{6} \times 2$ in number, but rarely present in full complement, the anterior being gradually cast; incisors in the upper jaw two (more or less prominent) at least in the male. Teeth of the complete series—at least of Trichechidae—M $\frac{6-6}{5-5}$, C $\frac{0}{1}$, I $\frac{2}{3} \times 2$; the upper incisors only persistent, the others as well as the canines being reabsorbed; molars successively increasing in size backwards.

1. Molars mostly with two or three roots each (generally three-rooted above and two-rooted below); and with crowns furnished with obtuse tubercles arranged in transverse yoke-like eminences, and in the posterior ones with an additional narrower tuberculated yoke behind the principal ones. (*Brandt.*)

HALITHERIIDAE. (LXI.)

2. Molars with simple hollow roots (not separated from the crowns); with crowns furnished with little prominent tubercles, few in number (in the anterior teeth simple or double, in the rest three or four) not forming yoke-like eminences, and early worn away and disappearing.

HALICORIDAE. (LXII.)

B. Teeth absent. (Intermaxillary lines with the apical portion produced and simulating incisorial teeth. Manducation is only effected by a very large palatine corneous plate, and by another opposed to it and covering the very large and elongated symphysis of the lower jaw.— *Brandt.*)

RHYTINIDAE. (LXIII.)

SUPER-FAMILY TRICHECOIDEA.

LX. TRICHECHIDAE.

Genus.

Trichechus Linn=*Manatus* Storr=*Oxystomus* Fischer.

SUPER-FAMILY HALICOROIDEA.

LXI. HALITHERIIDAE.

Extinct.

Halitherium Kaup, 1838=*Halianassa* v. Meyer, 1838.

Metaxytherium de Christol.
Halitherium Kaup.
Pugmeodon Kaup.

Fucotherium Kaup.
Pontotherium Kaup.
Cheirotherium Bruno.

LXII. HALICORIDAE.

Genus.

Halicore Illiger=*Dugungus* Tiedm=*Platystomus* Fisch.

LXIII. RHYTINIDAE.

Genus.

Extinct?

Rhytina Steller=*Stellerus* Desm.=*Nepus* Fisch.

SIRENIA? INCERTAE SEDIS.

Extinct.

Trachytherium Gervais.
Prorastomus Owen.
Anoplonassa Cope.
Hemicaulodon Cope.
Crassitherium Van Beneden.

VIII. CETE.

SUB-ORDERS.

I. Intermaxillaries expanded forwards, normally interposed between the maxillaries, and forming the terminal as well as anterior portion of the lateral margin of the upper jaw. Nasal apertures produced more or less forwards, and with the nasal bones freely projecting. Teeth of the intermaxillaries apparently in normal number (3+3), conic; of the maxillaries, 2- or 3-rooted.

ZEUGLODONTIA.

II. Intermaxillaries narrowed forwards, forming only the point of the upper jaw, and underlaid by the maxillaries, which form the entire lateral alveolar margins of the jaw. Nasal apertures far back, near the vertex, and with the nasal bones appressed. Teeth (when present) all single-rooted.

A. Teeth more or less persistent after birth. Upper jaw without baleen. Supramaxillary expanded backwards over the frontal bones, but not produced outwards in front of the orbits. Rami of lower jaw united by a symphyseal suture. Olfactory organ rudimentary or absent; the nasal bones appressed on the frontals and overlapped distally by the mesethmoid.

DENTICETE.

B. Teeth absorbed and disappearing before birth. Upper jaw provided with plates of baleen. Supramaxillary not expanded backwards over the frontal bones, but produced outwards in front of the orbits. Rami of lower jaw connected by fibrous tissue, and not by suture. Olfactory organ distinctly developed; the nasal bones projecting forwards, and not overlapped at their distal ends.

MYSTICETE.

ZEUGLODONTIA.

FAMILIES.

I. Parietal, frontal, and especially nasal bones elongated. Anterior nares open forwards. (Cope.)

BASILOSAURIDAE. (LXIV.)

II. Parietal, frontal, and especially nasal bones abbreviated. Anterior nares open far behind. (Cope.)

CYNORCIDAE. (LXV.)

LXIV. BASILOSAURIDAE.

Extinct genera.

Basilosaurus Harl.=*Zeuglodon* Owen=*Polyptychodon* Emmons=*Hydrarchos* Koch. *Durodon* Gibbes=*Pontogenus* Leidy.

LXV. CYNORCIDAE.

Extinct genera.

(Fide Copei.)

Portheodon Cope.
Squalodon Grat.=*Colophonodon* Leidy, Cope, 1867.
Cynorca Cope.
Delphinodon Leidy=*Squalodon* Cope, 1867.

Genera? incertae sedis.

Stenodon Van Ben.
Saurocetus Gibbes.

DENTICETE.

FAMILIES.

I. Rostrum of skull moderately prolonged, and terminating in a rounded or subangulated apex.

A. Head (generally) rostrated and attenuated, or ledge-like around the margin. Skull with the vertex produced forwards. Supraoccipital not projecting forwards laterally above the temporal fossæ. Frontals visible above only as elongated hook-shaped borders produced backwards around the maxillaries. (*Delphinoidea.*)

1. Lachrymal bones coalesced with the jugals.

    a. Costal cartilages not ossified. The tubercular and capitular articulations of the ribs blending together posteriorly. (Flower.)

        a1. Maxillary bones with crests null or little developed. Teeth in great part with a complete cingulum, or a distinct tubercle at the base of the crown. Eye moderate. External respiratory aperture transversely crescentiform.

                              INIIDAE. (LXVI.)

        a2. Maxillary bones with large bony incurved crests. Teeth without cingulum or tubercle. Eye rudimentary. External respiratory aperture longitudinal.

                      PLATANISTIDAE. (LXVII.)

    b. Costal cartilages firmly ossified. Posterior ribs losing their capitular articulation, and only uniting with the transverse processes of the vertebræ by the tubercle. (Flower.)

                      DELPHINIDAE. (LXVIII.)

2. Lachrymal bones distinct from the jugals.

    a. Costal cartilages not ossified. The hinder ribs losing their tubercular, and retaining their capitular articulation with the vertebræ. (Flower.)

                      ZIPHIIDAE. (LXIX.)

B. Head not rostrated or marginated; snout high towards the front and projecting beyond the mouth. Skull raised behind and retrorsely convex. Supraoccipital projecting forwards laterally to or beyond the vertical of the temporal fossæ. Frontals visible above as erect triangular or *retrorsely* falciform wedges between the maxillaries and supraoccipital. (*Physeteroidea.*)

                      PHYSETERIDAE. (LXX.)

II. Rostrum of skull prolonged into a slender, straight beak, the intermaxillary and maxillary bones forming a cylinder, bearing teeth on its proximal portion. (*Rhabdosteoidea.*)

                      RHABDOSTEIDAE. (LXXI.)

SUPER-FAMILY DELPHINOIDEA.

LXVI. INIIDAE.

Genus.

Inia D'Orb.

Extinct Iniidae?

Tretosphys Cope.
Zarhachis Cope.
Priscodelphinus Leidy.
Ixacanthus Cope.
Lophocetus Cope.

LXVII. PLATANISTIDAE.

Genus.

Platanista Cuv.

LXVIII. DELPHINIDAE.

SUB-FAMILIES.

I. Neck evident externally, the cervical region being attenuated. Frontal area longitudinally expanded an 1 little depressed. Postorbital process of frontal and zygomatic process of squamosal projecting outwards, and the latter enlarged and directed forwards. Maxillary with a crest and free margin over orbital region.

PONTOPORIINAE. (A.)

II. Neck not evident externally, the cervical region not being differentiated. Frontal area abbreviated and declivous. Postorbital process of frontal and zygomatic process of squamosal compressed, and the latter comparatively short and oblique. Maxillary with no supraorbital crest.

    1. Digits (second and third) not segmented into more than 5-6 phalanges, each.

      a. Cervical vertebræ all distinct.

DELPHINAPTERINAE. (B.)

      b. Cervical vertebræ more or less (2 to 7) consolidated.

DELPHININAE. (C.)

    2. Digits (second and third) segmented into numerous phalanges.

GLOBIOCEPHALINAE. (D.)

A. PONTOPORIINAE.

Pontoporia Gray=*Stenodelphis* Gerv.

B. DELPHINAPTERINAE.

Delphinapterus Lac., Lillj.=*Beluga* Gray.
Monodon Linn.

C. DELPHININAE.

Sotalia Gray.
Steno Gray.
Delphinus Linn.
Clymenia Gray.
Tursiops Gerv.=*Tursio* Gray.
Cephalorhynchus F. Cuv.=*Eutropia* Gray.
Lagenorhynchus Gray.

  Electra Gray.   *Feresa* Gray.
  Lucopleurus Gray.   *Lagenorhynchus* Gray.

Leucorhamphus Lillj.=*Delphinapterus* Gray (not Lac.)
Pseudorca Reinh.
Orca Gray

  Orca sensu strict.   *Ophysia* Gray.

Orcaella Gray.
Phocæna Gray.

  Phocæna sensu strict.   *Acanthodelphis* Gray.

Neomeris Gray.
Sagmatias Cope.

Globiocephalus Gray.
  Globiocephalus sensu strict.
  Grampus Gray.

Sphaerocephalus **Gray.**

LXIX. ZIPHIIDAE.

SUB-FAMILIES.

I. Maxillaries with no incurved lateral crests.

ZIPHIINAE. (A.)

II. Maxillaries with greatly developed incurved crests.

ANANARCINAE. (B.

A. ZIPHIINAE.

Ziphius Cuv.=*Epiodon* Gray.
  Epiodon Gray.   *Petroryhnchus* Gray.
Mesoplodon Gerv. = *Ziphius* Gray = *Heterodon* Blainv. 1816 (not Beauv.
  1800) = *Diodon* Less. = *Aodon* Less. = *Nodus* Wagl.
  Ziphius Gray.   *Dolichodon* Gray.
  Neoziphius Gray.   *Dioplodon* Gerv.
Berardius Duv.

B. ANARNACINAE.

Anarnacus Lac.=*Hyperoodon* Lac.=*Chenocetus* Eschr.
  Hyperoodon Gray.   *Lagenocetus* Gray.

Extinct Ziphiidae.

Choneziphius Duv.
Belemnoziphius Huxl.
Placoziphius Van Ben.
Ziphirostrum Van Ben.
Aporotus Du Bus.
Ziphiopsis Du Bus.
Rhinostodes Du Bus.

SUPER-FAMILY PHYSETEROIDEA.

LXX. PHYSETERIDAE.

SUB-FAMILIES.

I. Head very large, truncated in front. Blow-hole near the edge of the
snout. Cerebral cavity declining downwards. Jugal and zygomatic processes of squamosal connected.

PHYSETERINAE. (A.)

II. Head moderate, conic in front. Blow-hole frontal. Cerebral cavity
inclining upwards. Jugal and zygomatic processes of squamosal remote.

KOGIINAE. (B.)

A. PHYSETERINAE.

Physeter Linn.= *Catodon* Gray+*Physeter* Gray.
  Physeter sensu strict.   *Meganeuron* Gray.

B. KOGIINAE.

Kogia Gray=*Euphysetes* Wall.
Callignathus Gill.

Extinct Physeteridae?

Orycterocetus Leidy.
Ontocetus Leidy.

SUPER-FAMILY? RHABDOSTEOIDEA.
LXXI. RHABDOSTEIDAE.

Extinct genus.

Rhabdosteus Cope.

MYSTICETE.

I. Skull with the maxillary region slightly arched, and with short baleen plates. Rostrum broad at the base, gradually tapering, depressed. Frontals with the orbital processes moderately prolonged, broad, and flat on the upper surface. (Supramaxillary bones with the posterior margin deeply excavated.) Tympanic bones elongated, ovoid. Lower jaw with the coronoid process more or less developed. Cervical vertebræ in whole or in part separated. Manus narrow, with four digits (first wanting). (Flower.)

BALAENOPTERIDAE. (LXXII.)

II. Skull with the maxillary region greatly arched, and with long, narrow baleen plates. Rostrum narrow and compressed at the base. Frontals with the orbital processes much prolonged, and extremely narrow and rounded on the upper surface. (Supramaxillary bones with the posterior margins entire.) Tympanic bones broad, rhomboid. Lower jaw with the coronoid processes scarcely perceptible. Cervical vertebræ coalesced. Manus broad, with five digits. (Flower.)

BALAENIDAE. (LXXIII.)

LXXII. BALAENOPTERIDAE.
SUB-FAMILIES.

I. Throat not plicated. Dorsal fin null.

AGAPHELINAE. (A.)

II. Throat longitudinally plicated. Dorsal fin developed.

  A. Frontal with the orbital processes much narrowed externally. (Flower.) Manus very long, with the four digits segmented into many phalanges. Dorsal fin hump-like.

MEGAPTERINAE. (B.)

  B. Frontal processes with the orbital processes nearly as broad at the outer extremity as the base, or somewhat narrowed. (Flower.) Manus moderate, with the four digits having each not more than six phalanges. Dorsal fin high, erect, falcate or subfalcate.

BALAENOPTERINAE. (C.)

A. AGAPHELINAE.

Agaphelus Cope.
Rhachianectes Cope.

February, 1872.

Megapatera Gray.
Poescopia Gray.
Eschrichtius Gray.

C. BALAENOPTERINAE.

§. 1.

Physalus Gray.
  Benedenia Gray.   *Physalus* Gray.
  Cuvierius Gray.
Sibbaldius Gray, 1866=*Flowerius* Lillj. 1867.
Rudolphius Gray (s. g.), 1866=*Sibbaldius* Lillj. 1867.

§. 2.

Balaenoptera Lac.
  Balaenoptera sensu strict.   *Swinhoia* Gray.

Extinct genera incertae sedis.

Cetotherium Brandt.
Plesiocetus Van Ben and Gerv.

LXXIII. BALAENIDAE.

Genera.

(*Fide Gray.*)

Balaena Linn.
Neobalaena Gray.
Eubalaena Gray.
Hunterius Gray.
Caperea Gray.
Macleayius Gray.

Extinct Balaenidae?

Palaeocetus Seeley.

www.ingramcontent.com/pod-product-compliance
Lightning Source LLC
Chambersburg PA
CBHW021827190326
41518CB00007B/768